3rd International Conference on Smart Materials Applications (ICSMA 2020)

IOP Conference Series: Materials Science and Engineering
Volume 855

Seoul, South Korea
13 - 16 January 2020

ISBN: 978-1-7138-2685-9
ISSN: 1757-8981

Printed from e-media with permission by:

Curran Associates, Inc.
57 Morehouse Lane
Red Hook, NY 12571

Additional copies of this publication are available from:

Curran Associates, Inc.
57 Morehouse Lane
Red Hook, NY 12571 USA
Phone: 845-758-0400
Fax: 845-758-2633
Email: curran@proceedings.com
Web: www.proceedings.com

TABLE OF CONTENTS

IOP Conf. Series: Materials Science and Engineering **855** (2020) 011001 doi:10.1088/1757-899X/855/1/011001

PREFACE

It is with deep satisfaction that I write this Foreword to the Proceedings of 2020 the 4th International Conference on Smart Materials Applications (ICSMA 2020) held at Yonsei University, Seoul, South Korea during January 13-16, 2020.

The ICSMA 2020 provided a premier interdisciplinary platform for scientists, researchers, industry leaders, engineers and educators throughout the world to present and discuss the most recent innovations, trends, concerns, as well as practical challenges encountered, and streamline solutions in the fields of Smart Materials. Conference has received more than 50 submissions from all over the world and after several rounds of review procedure by the technical committees. Some excellent papers have been received to get published in the conference proceedings. And the authors also have been invited to make the presentation in the conference to share their research achievements in 8 oral sessions and 1 poster session.

The conference particularly encouraged the interaction of research students and developing academics with the more established academic community in an informal setting to present and to discuss new and current work. Their contributions helped to make the conference as outstanding as it has been. The papers contributed the most recent scientific knowledge known in the field of Materials Science and Engineering, Materials Properties, Measuring Methods and Applications.

In addition to the contributed papers, three plenary and two invited presentations were given by:

Prof. Michael D. Guiver, Tianjin University, China
Prof. Ki Bong Lee, Korea University, South Korea
Prof. Xiaohong Zhu, Sichuan University, China
Prof. C.Q. RU, University of Alberta, Canada
Prof. Juan C. Suárez-Bermejo, Technical University of Madrid (UPM), Spain

These Proceedings will furnish the scientists of the world with an excellent reference book. I trust also that this will be an impetus to stimulate further study and research in all these areas.
We thank all authors and participants for their contributions.

Much of the credit of the success of the conference is due to topic coordinators who have devoted their expertise and experience in promoting and in general co-ordination of the activities for the organization and operation of the conference. The coordinators of various session topics have devoted a considerable time and energy in soliciting papers from relevant researchers for presentation at the conference.

We would like to thank everyone who was part of this conference and would like to see you again at ICSMA 2021.

Conference Chairs
On behalf of the ICSMA Conference Committee
Prof. Xiaohong Zhu, Sichuan University, China
January 13-16, 2020

IOP Conf. Series: Materials Science and Engineering **855** (2020) 011002 doi:10.1088/1757-899X/855/1/011002

CONFERENCE COMMITTEES

Conference Committee Chairs

Prof. Jong Hak Kim, Yonsei University, South Korea

Prof. Xiaohong Zhu, Sichuan University, China

Local Chair

Prof. Jung Tae Park, Konkuk University, Korea

Program Committee Chairs

Prof. Ki Bong Lee, Korea University, South Korea

Prof. Sunwoong Choi, Hannam University, South Korea

Prof. Yu-Liang Chen, National Defense University, Taiwan

Technical Committee Chairs

Prof. Hu Yan, Zhengzhou University, China

Prof. Yu-Chieh Cheng, National Taipei University of Technology (NTUT), Taiwan

Prof. Wen-Jauh CHEN, National Yunlin University of Science and Technology, Taiwan

Prof. Shyh Ming Kuo, I-Shou University, Taiwan

Prof. Sapuan S.M., Universiti Putra Malaysia, Malaysia

Assoc. Prof. Abdul Hai Alami, University of Sharjah, United Arab Emirates

Assoc. Prof. Babak Salamatinia, Monash University Malaysia, Malaysia

Assoc.Prof. Kowit Piyamongkala, King Mongkut's University of Technology North Bangkok, Thailand

Assoc. Prof. Toempong Phetchakul, King Mongkut's Institute of Technology Ladkrabang, Thailand

Assoc. Prof. Jose Isagani B. Janairo, De La Salle University, Philippines

Assoc. Prof. Terence Tumolva, University of the Philippines, Diliman, Philippines

Assoc. Prof. Sutanu Samanta, North Eastern Regional Institute of Science and Technology (NERIST), India

Assoc. Prof. Harish Narayana, Ramiah Institute of Technology, India

Assoc. Prof. Hosein Naderpour, Semnan University, Iran

Asst. Prof. Peng Hao Wang, Purdue University, USA

Asst. Prof. Chih-Ling Huang, Kaohsiung Medical University, Taiwan

Asst. Prof. Pat Sooksaen, Silpakorn University, Thailand

Asst. Prof. Wornchanok Chaiyasoonthorn, King Mongkut's Institute of Technology Ladkrabang, Thailand

IOP Conf. Series: Materials Science and Engineering **855** (2020) 011002 doi:10.1088/1757-899X/855/1/011002

First name	Surname
Qi Hwa	Ng
Bochao	Wang
Carmen	Mihai
Nabila	Alifani
Qi Hwa	Ng
Nuttapol	Lerkkasemsan
Suchánek	Jan
Toempong	Phetchakul
Tanawit	Srisantirut
Sergey	Inozemtcev
Wael	Zaki
Dostál	Michal

Institution
Universiti Malaysia Perlis, Malaysia
KTH Royal Institute of Technology, The Marcus Wallenberg Laboratory for
The Research and Development Institute for Textiles and Leather, Romania
Universitas Indonesia, Indonesia
Universiti Malaysia Perlis, Malaysia
Department of Chemical Engineering, Faculty of Engineering, King
J. Heyrovsky Institute of Physical Chemistry, Academy of Sciences of the
Faculty of Engineering, King Mongkut's Institute of Technology
Faculty of Engineering, King Mongkut's Institute of Technology
REC "Nanomaterials and Nanotechnology", NRU MGSU, 129337
Khalifa University, United Arab Emirates
J. Heyrovsky Institute of Physical Chemistry, Academy of Sciences of the

E-mail address
qhng@unimap.edu.my
bochao@kth.se
carmen.mihai@incdtp.ro
nabila.alifani@yahoo.com
qhng@unimap.edu.my
Nuttapol.le@kmitl.ac.th
zdenek.zelinger@jh-inst.cas.cz
toempong.ph@kmitl.ac.th
60601070@kmitl.ac.th
inozemtsevss@mail.ru
wael.zaki@ku.ac.ae
zdenek.zelinger@jh-inst.cas.cz

IOP Conf. Series: Materials Science and Engineering **855** (2020) 011004 doi:10.1088/1757-899X/855/1/011004

Peer review statement

All papers published in this volume of *IOP Conference Series: Materials Science and Engineering* have been peer reviewed through processes administered by the proceedings Editors. Reviews were conducted by expert referees to the professional and scientific standards expected of a proceedings journal published by IOP Publishing.

IOP Conf. Series: Materials Science and Engineering **855** (2020) 012001 doi:10.1088/1757-899X/855/1/012001

Screening of Process Parameters to Produce Xylanase from Aspergillus niger for Secondary Bioethanol Production

M Tarrsini[1], Q H Ng[1,*], Y P Teoh[2,], B Kunasundari[1], W U Ang[1], S H Shuit[3], Z X Ooi[4] and P Y Hoo[1]**

[1]Department of Chemical Engineering Technology, Faculty of Engineering Technology, Universiti Malaysia Perlis, Kampus UniCITI Alam, Sungai Chuchuh, Padang Besar, 02100, Perlis, Malaysia.
[2]Department of Petrochemical Engineering, Faculty of Engineering and Green Technology, Universiti Tunku Abdul Rahman, Jalan Universiti, Kampar, Perak, Malaysia
[3]Department of Chemical Engineering, Lee Kong Chian Faculty of Engineering and Science, Universiti Tunku Abdul Rahman, Sungai Long Campus, Jalan Sungai Long, Bandar Sungai Long, Cheras, 43000, Kajang, Selangor, Malaysia.
[4]Department of Chemical Science, Faculty of Science, Universiti Tunku Abdul Rahman, Jalan Universiti, Bandar Barat, 31900, Kampar, Perak, Malaysia.

Email: qhng@unimap.edu.my*; teohyp@utar.edu.my**

Abstract. In recent years, the biotechnological use of xylanases has grown remarkably. Xylanase is a hydrolytic enzyme with a broad industrial application. In specific, xylanase can convert xylan into xylose, a fermentable sugar source for secondary bioethanol production. The objective on this study is to investigate the significance of different parameter effects for an efficient xylanase production from *Aspergillus niger* (*A. niger*). In this study, four factors: incubation temperature, medium pH, incubation time, and agitation speed were screened by performing One-factor-at-a-time (OFAT) analysis. Xylanase production with the maximal enzyme activity was successfully obtained from OFAT analysis under condition of 32°C, pH 5.0, 5 days, and 150 rpm.

1. Introduction

Agricultural and forest industries produce hundred-million tons of lignocellulose wastes yearly, which contain approximately 30–40% of hemicellulose in dry weight [1]. Due to improper handling and lack of technical know-how, these waste substances could be a cause of environmental pollution [2]. Many processing industries which are generating agro-waste substances or by-products are struggling to challenge for their conversion into value-added products. Therefore, enzyme, carbo-hydrolase is the excellent alternative for conversion of cellulose and hemicellulose into value added products [3]. From many research developed, the use of xylanases promised such processes to be run with fewer chemicals, under less harsh conditions, and with less severe side reactions [4]. Xylanases can be included in bioconversion of lignocellulosic material such as wheat bran, rice husk, rice straw, sugarcane bagasse and other agro-wastes into fermentative products [5]. This process can occur under simultaneous saccharification and fermentation. This application can recycle the wasted biomass into useful energy such as bioethanol. Thus, it is indirectly minimizing the amount of waste from industry [6].

Yet, the overly priced commercial xylanases are known to be a bottleneck for the secondary bioethanol production. Furthermore, the cost of enzyme production is one of the predominant factors determining the economics of a process [7]. Reducing the costs of enzyme production by optimization of the fermentation medium and the process is the goal of basic research for industrial application. The microorganisms used for xylanase production must be non-pathogenic, capable of producing high levels of extracellular enzymes and most important is it can be cultivated very easily. Studied data have shown that xylanases can be produced by a diverse variety of microorganism [8]. From an industrial point of view, filamentous fungi such as *A. niger* is a preferred producer of extra-cellular xylanase enzyme as compared to bacteria and yeast [9]. Moreover, the mentioned fungi also produce numerous auxiliary enzymes, which are fundamental for debranching of the substituted xylan [9]. However, research on the screening condition of *A. niger* in submerged state fermentation to produce maximum xylanase is limited. Hence, it is important to study on the screening process for a better knowledge on the factors that have a significant role in maximizing the xylanase production from *A. niger*.

2. Material and Methods

2.1. Materials
All the chemicals used in this research work were analytical grade. These chemicals were purchased from Sigma-Aldrich, St. Louis, MO USA except otherwise stated.

2.2. Microorganism Maintenance
Fungal strain identified as *A. niger* was obtained from Universiti Tunku Abdul Rahman, (UTAR). The strain was maintained on Potato Dextrose Agar (PDA) slants by monthly transfer and was stored at 4°C.

2.3. Inoculum Development
A 10 mL of 1% (v/v) Tween-80-sterile distilled water was added to a 5 days old pure *A. niger* culture on agar plate and the agar surface was rubbed with a sterile L-shaped spreader to obtain spore suspension. The suspension was sieved through Whatman no.1 paper to remove mycelia. After appropriate dilutions, inoculum sizes (2 x 10^6 spores/mL) were determined by direct microscopic observation using hemocytometer.

2.4. Submerged Fermentation (SmF)
Xylanase biosynthesis were performed in a 250 mL Erlenmeyer flask using SmF in a production medium containing 100 mL of Mandels and Sternburg's basal medium and 5.0 g/L of xylan as carbon source. Composition of the growth medium includes: 2.0 g/L NH_4NO_3, 2.0 g/L K_2HPO_4, 1.0 g/L $MgSO_4.7H_2O$, 0.3 g/L $CaCl_2$, and 0.012 g/L trace elements: 5.0 mg/L $FeSO_4.7H_2O$, 1.6 mg/L $MnSO_4.4H_2O$, 3.45 mg/L $ZnSO_4.7H_2O$ and 2.0 mg/L $CoCl_2.6H_2O$ and 0.1% (v/v) Tween 80. The medium was sterilized at 121°C for 15 mins. Aliquot of 1mL of standardized *A. niger* inoculum (2 x 10^6 spores/mL) was aseptically inoculated into the medium and fermentation was allowed to take place at different selected incubation temperature, incubation period and agitation speed. Crude enzyme was extracted with 0.05 M citrate buffer pH 4.8 (50 mL). After the addition of the buffer, the whole flask was kept in a rotary incubator at 150 rpm (32°C) for 15 min. The cell was filtered through Whatman no.1 filter paper to extract the filtrate which was then centrifuged at 10,000 rpm for 15 min at 4°C. The filter paper was dried at 60°C for three consecutive days in an oven (Memmert, Germany). Mycelial Dry Mass per Volume (MDMV) was determined by weighing the filter paper on an analytical balance (Sartorius, Germany). The supernatant was collected as the crude enzyme for xylanase activity analysis [10].

IOP Conf. Series: Materials Science and Engineering **855** (2020) 012001 doi:10.1088/1757-899X/855/1/012001

2.5. Screening Factors Affecting Xylanase Production

Factors that affect xylanase production was screened by performing one-factor-at-time (OFAT) experiments, which vary only one factor or a variable at a time while keeping others fixed. Four parameters that manipulated were incubation temperature, medium pH, agitation speed, and incubation time. Each parameter was varied into a few sets of experiments. The pH used in this study was 3.0, 5.0, 7.0 and 9.0. Meanwhile, the temperatures used were at 25, 28, 32, and 37°C respectively for the screening study. Medium agitation speed used were 50, 100, 150, and 200 rpm. The culture incubation time was varied from day intervals. The temperature, pH, incubation period and agitation speed were kept constant at 32°C, pH 5, 5 days and 150 rpm, respectively when not set as a manipulative variable.

3. Result and Discussion

Screening analysis was done via OFAT technique to determine the parameters that significantly influenced the xylanase production. The effect of temperature on xylanase production in the range of 25 to 37°C was examined (Figure 1(a)). The temperature of the medium has governed role in a fermentation process. An extremely low or high temperature setting could denature microbial growth and enzyme synthesis [11]. Figure 1(a) indicates that mycelial dry mass per volume (MDMV) of *A. niger* increased from 25°C, until optimal MDMV of 0.270 mg/mL was attained at 32°C. In common, the temperature with the highest growth of fungi results the maximal xylanase production [12]. The enzyme activity displayed a similar trend as the MDMV toward the incubation temperature. The highest enzyme activity (0.524 U/mL) was detected at the optimal *A. niger* growth temperature of 32°C.

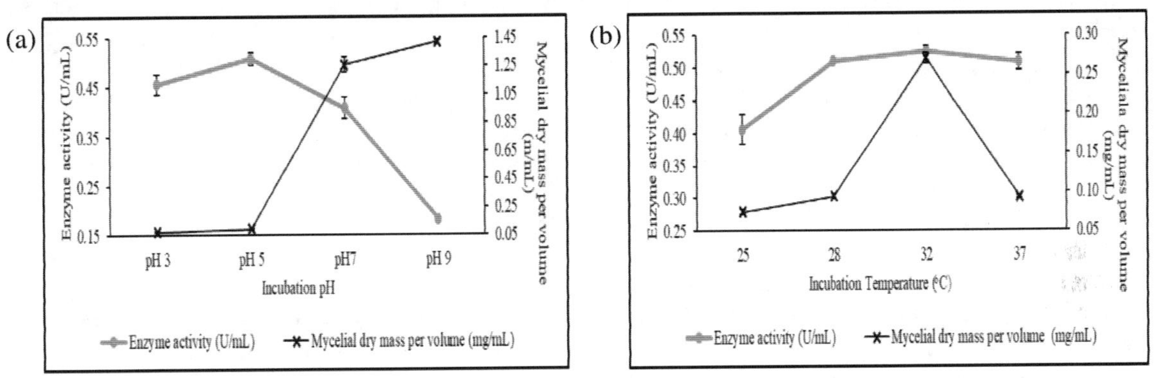

Figure 1: (a) Effect of incubation temperature on xylanase production and (b) Effect of incubation pH on xylanase production

pH is a key parameter in most biological processes and have various effects on the biomass, including on the pellet formation, the growth of biomass; synthesis of extracellular enzyme, its secretion and the stability of the enzymatic system [13]. The result of different medium pH on enzyme production by *A. niger* is illustrated in Figure 1(b). From Figure 1(b), is can be concluded that pH 5.0 was the most favourable xylanase production condition as highest enzyme activity (0.508 U/mL) was recorded. It can be also observed that MDMV at alkaline pH (7 and 9) was greater than at pH 5 but this response did not boost the enzyme activity. This was due to the fact that an alkaline fermentation media can affect enzyme stability, which caused enzyme denaturation occurred [9].

Figure 2(a) illustrates the effect of the incubation time on the xylanase production. The enzyme activity was at its peak of 0.675 U/mL with maximum MDMV (0.083 mg/mL) when the incubation time was prolonged to 5 days. This due to the fact that *A. niger* achieved its maximum growth in the stationary phase, which is approximately after 5 days [11]. Subsequently, the production of xylanase was at optimum state. There was a sharp decrement in MDMV at the incubation was continued for 7

IOP Conf. Series: Materials Science and Engineering **855** (2020) 012001 doi:10.1088/1757-899X/855/1/012001

and 9 days as the growth of the *A. niger* was entering death phase whereby there will be poor availability of oxygen and toxic accumulation in the fermentation media [14].

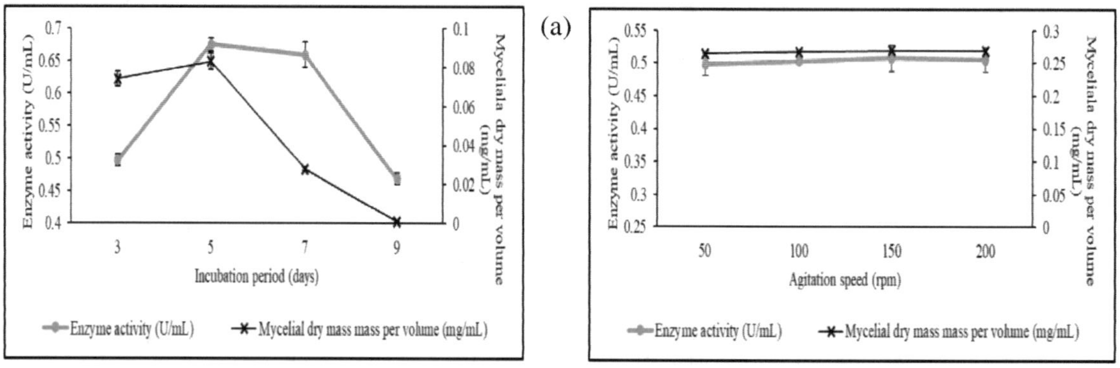

Figure 2: Effect of incubation time on xylanase production and (b) Effect of agitation speed on xylanase production

Last but least, the effect of agitation speed was selected to be investigated. This is because the agitation speed has an influence on the mixing extent and nutrient availability in the xylanase production media was investigated [15]. Figure 2(b) shows that the effect of the agitation speed on the xylanase production. From the Figure 2(b), it can be seen that the enzyme activity and MDMV increased with the increase in agitation until the optimal reading of 0.508 U/mL and 0.270 mg/mL, respectively was achieved at 150 rpm. Additionally, a slight decrease was observed for the MDMV (0.269 mg/mL) and enzyme activity (0.505 mg/mL) when the agitation speed was further raised to 200 rpm. The adequate aeration and nutrient such as dissolved oxygen (DO) by increased agitation speed might cause the fungus to grow well, but shear forces and cell damage due to collision among the fungal pellets finally had a negative effect on enzyme production [16].

4. Conclusion

This research work focused on evaluation of the effect of process parameters for xylanase production. There were four main parameters that affect the xylanase yield from *A. niger* in production medium which were incubation temperature, medium pH, incubation period, and agitation speed. The highest enzymatic activity was recorded at 32 °C (0.525 U/mL), pH 5.0 (0.508 U/mL), 5 days (0.675 U/mL), and 150 rpm (0.508 U/mL), respectively. Finally, it can be concluded that through OFAT analysis, incubation temperature and pH were the two most influential parameters on the xylanase production process. The notable deviation in enzymatic activity within the range set for incubation temperature (25, 28, 32, and 37°C) and pH (3, 5, 7 and 9) contributes the significance of these parameters. Nevertheless, the results obtained from OFAT analysis was only a surface guidance to individually recognize the specific optimal enzymatic activity and significance of each analyzed parameter. Hence, in future, optimization based research work should be conducted for the xylanase production through a Response Surface Methodology (RSM). With this, the combined interaction effect of the parameters and an overall optimized condition for maximized xylanase production to be aided in xylan to bioethanol conversion could be identified.

5. Reference

[1] Joshi B, Bhatt M R, Sharma D, Joshi J, Malla R and Sreerama L 2011 Lignocellulosic ethanol production *Biotechnology and Molecular Biology Review* **6** 172–82

[2] Fitria, Ruan H, Fransen S C, Carter A H, Tao H and Yang B 2019 Selecting winter wheat straw for cellulosic ethanol production in the Pacific Northwest, U. S. A *Biomass and Bioenergy* **123** 59–69

[3] Meddeb-mouelhi F, Kelly J and Beauregard M 2014 A comparison of plate assay methods for detecting extracellular cellulase and xylanase activity *Enzyme and Microbial Technology* **66** 16–9

[4] Oliveira P, Cavalieri N, Guimaraes D A, Carvalho S De, Betini J H, Marchetti C R, Fonseca F, Lourdes M D, Moraes T D, Marques M R and Giannesi C 2015 Production of cellulase-free xylanase by *Aspergillus flavus* : Effect of polyols on the thermostability and its application on cellulose pulp biobleaching *African Journal of Biotechnology Full* **14** 3368–73

[5] Ncube T, Howard R L, Abotsi E K, Rensburg E L J Van and Ncube I 2012 *Jatropha curcas* seed cake as substrate for production of xylanase and cellulase by *Aspergillus niger* FGSCA733 in solid-state fermentation *Industrial Crops & Products* **37** 118–23

[6] Song Y, Jin E, Song C, Hoon C, Park B and Bae H 2019 A strategy for sequential fermentation by *Saccharomyces cerevisiae* and *Pichia stipitis* in bioethanol production from hardwoods *Renewable Energy* **139** 1281–9

[7] Singh G, Singh H, Kaur S, Bansal S and Kaur S 2011 Value-addition of agricultural wastes for augmented cellulase and xylanase production through solid-state tray fermentation employing mixed-culture of fungi *Industrial Crops & Products* **34** 1160–7

[8] Chadha B S, Kaur B, Basotra N, Tsang A and Pandey A 2019 Thermostable xylanases from thermophilic fungi and bacteria : Current perspective *Bioresource Technology* **277** 195–203

[9] Modi H A, Acharya P B, Chabhadiya S B, Shah A J, Acharya D K, Acharya, P. B. Chabhadiya S B, Shah A J and Acharya D K 2011 Process optimization for xylanase production by *Aspergillus niger* in solid state fermentation *International Journal of Biotechnology and Biosciences* **1** 423–30

[10] Bailey M J, Biely P and Kaisa P 1992 Interlaboratory testing of methods for assay of xylanase activity *Journal of Biotechnology* **23** 257–70

[11] Ajijolakewu A K, Leh C P, Nadiah W W A and Lee C K 2017 Optimization of production conditions for xylanase production by newly isolated strain *Aspergillus niger* through solid state fermentation of oil palm empty fruit bunches *Biocatalysis and Agricultural Biotechnology* **11** 239–47

[12] Kheng P P and Omar I C 2005 Xylanase production by a local fungal isolate , *Aspergillus niger* USM AI 1 via solid state fermentation using palm kernel cake (PKC) as substrate *Songklanakarin Journal of Science Technology* **27** 325–36

[13] Rostami K and Mirdamadi S 2008 Some studies of a-Amylase Production Using *Aspergillus oryzae Pakistan Journal of Biological Sciences* **11** 2553–9

[14] Gliseida B, Melgar Z, Vanessa F, Assis S De, Coutinho L, Fanti S C and Sette L D 2013 Growth curves of filamentous fungi for utilization inbiocatalytic reduction of cyclohexanones *Global Journal of Science Frontier Research Chemistry* **13** 13–9

[15] Shankar U, Uday P, Choudhury P, Kanti T, Uday U S P and Choudhury, P. Bandyopadhyay, T. K. Bhunia B 2016 Classification, mode of action and production strategy of xylanase and its application for biofuel production from water hyacinth *International Journal of Biological Macromolecules* **82** 1041–54

[16] Ibrahim D and Lim S 2015 Effect of agitation speed on the morphology of *Aspergillus niger* HFD5A-1 hyphae and its pectinase production in submerged fermentation *World Journal of Biological Chemistry* **26** 265–71

Acknowledgement

The author would like to acknowledge the support from the Fundamental Research Grant Scheme (FRGS) under a grant number of FRGS/1/2017/TK02/UNIMAP/02/7 from the Ministry of Education Malaysia. Furthermore, the authors are also grateful to Universiti Tunku Abdul Rahman for the financial aid through the UTAR Research Fund (UTARRF) with a project number IPSR/RMC/UTARRF/2019-C1/T05.

IOP Conf. Series: Materials Science and Engineering **855** (2020) 012002 doi:10.1088/1757-899X/855/1/012002

One dimensional constitutive model of isotropic magneto-sensitive rubber under shear deformation with amplitude, frequency and magnetic dependency

B Wang and L Kari

KTH Royal Institute of Technology, The Marcus Wallenberg Laboratory for Sound and Vibration Research (MWL), Teknikringen 8, 100 44 Stockholm, Sweden

bochao@kth.se

Abstract. Modelling the amplitude, frequency and magnetic dependency of magneto-sensitive (MS) rubber in one dimensional case under shear deformation is proposed in this paper. MS rubber is a kind of smart materials which mainly consists of magnetizable particles and rubber matrix. The modulus of MS rubber changes rapidly after the applying of magnetic field which is referred as the magnetic dependency of MS rubber. Besides the magnetic dependency, there is an amplitude and frequency dependency of MS rubber. Specifically, the shear modulus of MS rubber under larger strain magnitude is smaller than the corresponding modulus under small strain and the shear modulus of MS rubber increases with increasing of frequency. A one-dimensional constitutive model consists of a fractional derivative element in parallel with a nonlinear kinematic elastoplastic model to depict the amplitude, frequency and magnetic dependency of MS rubber is developed in this paper. After comparison with the test result, it found that with eight parameters the amplitude, frequency and magnetic dependency can be depict precisely by the model developed. It is helpful for the predicting the dynamic behaviours of MS rubber-based vibration devices in the design phase and promotes MS rubber's application in sound and vibration area.

1. Introduction

Magneto-sensitive (MS) rubber is a solid substitute for magnetorheological fluids with main components of magnetizable particles and non-magnetizable matrix. Due to the stiffness changing capacity, there are many researches of MS rubber's possible application in vibration reduction area. For instance, Deng and Gong [1] developed a stiffness adaptable vibration absorber based on MS rubber. Test results showed that the vibration of the main structure can be reduced substantially by changing the stiffness of the MS rubber isolator to trace the frequency of the external harmonic loading. Wang and Kari [2] developed a MS rubber-based vibration isolator and change its stiffness by fuzzy logical controller. The simulation results revealed that compared with the traditional rubber-based vibration isolator without variable stiffness, the vibration isolation effect of MS rubber isolator is obviously enhanced. Similar results can be found in the articles by Jung [3], Sun [4] and Yang [5].

It is foreseeable that with the improvement of noise control requirements and progress in material science, the investigation of MS rubber will expand. A constitutive model which depicts the mechanical properties of MS rubber precisely is of great importance. There were many constitutive models of MS rubber, like the magnetic dipole-based models [6-7] and continuum mechanics-based models [8-9]. The drawback for the models mentioned is that only the magnetic-elastic and

IOP Conf. Series: Materials Science and Engineering **855** (2020) 012002 doi:10.1088/1757-899X/855/1/012002

viscoelastic behaviours are simulated. However, the dynamic shear test by Lejon and Kari [10] shows except the magnetic-elastic and viscoelastic behaviours there is an amplitude dependency of mechanical behaviours of MS rubber. To be specific, the shear modulus of MS rubber decreases with the increasing of strain amplitude. However, the amplitude dependency is often neglected for the constitutive modelling of MS rubber which is unfavorable for the prediction of MS rubber's mechanical properties during application. Therefore, a constitutive model with amplitude dependency of MS rubber is needed and necessary.

Normally the MS rubber vibration reduction devices [2-5] is constructed as follows: MS rubber is imbedded by two iron plates and works in a translational shear mode. A one-dimensional constitutive model to determine the shear stress strain relation is developed in this paper. The advantage of this model is that only eight parameters are needed to depict the amplitude, frequency and magnetic dependency of MS rubber which is helpful for the prediction of dynamic behaviours of MS rubber-based vibration reduction devices.

2. Constitutive model

The basic idea of the constitutive model is that the total shear stress is additively decomposed into viscoelastic τ^{ve} and elastoplastic stress τ^{ep} and the rheological configuration is visualized in figure 1. The meaning of material parameters in figure 1 will explained in the following section. Mathematically, the additive decomposition of the total shear stress is

$$\tau = \tau^{\mathrm{ve}} + \tau^{\mathrm{ep}} . \tag{1}$$

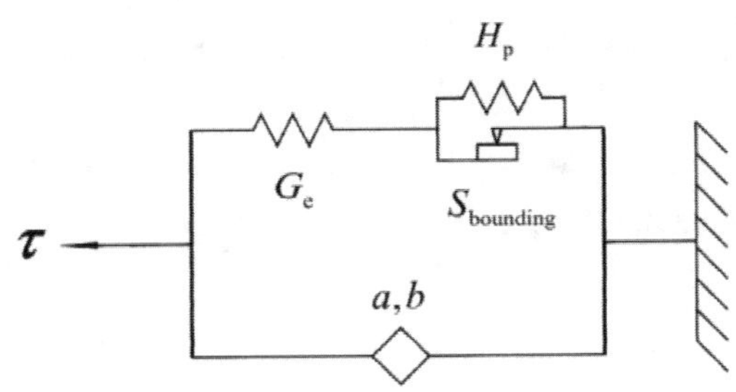

Figure 1. Rheological configuration of the constitutive model.

It is assumed that τ^{ve} is not affected by the magnetic field since physically the viscoelastic behaviour is mainly caused by the non-magnetizable rubber matrix. The measurement result that the loss factor of MS rubber is a relatively constant value with respect to the magnetic field supporting this assumption. The constitutive equation for the shear strain γ and τ^{ve} is depicted by a fractional derivative viscoelastic model which is

$$\tau^{\mathrm{ve}}(t) = \frac{b}{\Gamma(1-a)} \frac{\mathrm{d}}{\mathrm{d}t} \int_0^t \frac{\gamma(s)}{(t-s)^a} \mathrm{d}s , \tag{2}$$

where Γ is the gamma function and a and b are parameters for fractional derivative model. The restriction for a is that $a \in [0,1]$. Compared with Generalized Maxwell model, only two parameters are needed for fractional derivative model to depict the viscoelastic behaviours with a higher accuracy for rubbers [11]

The elastoplastic stress is determined by a bounding surface nonlinear kinematic hardening model in series with an elastic spring. Symbols G_e, S_bounding and H_p are the parameters for the elastoplastic model. Measurement result reveals that the shear modulus and the magnetic induced shear modulus under a larger strain amplitude is smaller than the corresponding values under small strain amplitude. The conclusion that the amplitude dependency and magnetic dependency are correlated with each

IOP Conf. Series: Materials Science and Engineering **855** (2020) 012002 doi:10.1088/1757-899X/855/1/012002

other can be drawn. Therefore, magnetic dependency is introduced for the parameters G_e, S_{bounding} and H_p. With considering the magnetic saturation of MS rubber, a hyperbolic tangent function is used

$$G_e = G_{e0}\left[1+\kappa_1 \tanh\left(\frac{I_h}{I_s}\right)\right], \tag{3}$$

$$S_{\text{bounding}} = S_{\text{bounding0}}\left[1+\kappa_2 \tanh\left(\frac{I_h}{I_s}\right)\right], \tag{4}$$

and

$$H_p = H_{p0}\left[1+\kappa_3 \tanh\left(\frac{I_h}{I_s}\right)\right], \tag{5}$$

where G_{e0}, $S_{\text{bounding0}}$ and H_{p0} are parameters under zero magnetic field. Symbols κ_1, κ_2 and κ_3 represent the magnetic field influencing factor. $I_h = H \cdot H$ with H the magnetic field intensity is the magnetic field invariant. I_s representing the magnetic saturation of MS rubber.

The bounding surface model within the elastoplastic model [12] is defined

$$\Phi = \left|\tau^{\text{ep,image}} - \beta\right| - S_{\text{bounding}} = 0, \tag{6}$$

where $\tau^{\text{ep,image}}$ is the image stress on the bounding surface which is determined by the kinematic rule and the extension direction of τ^{ep}. Let $[t_{n-1}, t_n]$ be the time interval of interest. Symbols with subscripts n and $n-1$ represent the quantities at the previous and current time step, respectively. Initially the plastic strain is locked at the current time step and the trial stress can be determined by

$$\tau^{\text{ep,trial}} = G_e\left(\gamma_n - \gamma_{n-1}^{p}\right), \tag{7}$$

The extension direction for τ^{ep} is obtained as

$$m = \frac{\tau^{\text{ep,trial}} - \tau_{n-1}^{\text{ep}}}{\left|\tau^{\text{ep,trial}} - \tau_{n-1}^{\text{ep}}\right|}. \tag{8}$$

According to Mróz's kinematic rule [13], $\tau^{\text{ep,image}}$ is determined as the intersection between m and bounding surface in equation (6). Mathematically, it is

$$\tau^{\text{ep,image}} = \tau^{\text{ep,trial}} + m\delta, \tag{9}$$

where $\delta = \left\{\left[m:\left(\tau^{\text{ep,trial}} - \beta_{n-1}\right)\right]^2 + S_{\text{bounding}}^2 - \left(\tau^{\text{ep,trial}} - \beta_{n-1}\right)^2\right\}^{1/2} - m:\left(\tau^{\text{ep,trial}} - \beta_{n-1}\right)$. The normal direction from the center of the bounding surface β to $\tau^{\text{ep,image}}$ is

$$N_n = \frac{\tau^{\text{ep,image}} - \beta_{n-1}}{\left|\tau^{\text{ep,image}} - \beta_{n-1}\right|}. \tag{10}$$

After obtaining $\tau^{\text{ep,image}}$, according to ref. [14] the plastic modulus H_s which used to determine τ^{ep} is obtained

$$H_s = H_p\left(\frac{\delta}{\delta_{\text{in}} - \delta}\right), \tag{11}$$

where δ_{in} is the maximum distance between τ^{ep} and $\tau^{\text{ep,image}}$ during a smooth loading history. Initially, $\delta_{\text{in}} = S_{\text{bounding}}$. The value of δ_{in} updates when a drastic change of loading occurs. The updating criteria is if $m: N_{n-1} < 0$ or $\delta > \delta_{\text{in}}$, then $\delta_{\text{in}} = \delta$. The plastic hardening rule to determine the plastic multiplier λ for bounding surface model is

$$H_{\text{s}}\lambda = \left(\tau_n^{\text{ep}} - \tau_{n-1}^{\text{ep}}\right) : N_n. \tag{12}$$

By equation (7) and the normality condition for associated plasticity that $\gamma_n^{\text{p}} = \lambda N_n$, the plastic multiplier λ can be determined. Which is

$$\lambda = \frac{\left(\tau^{\text{ep,trial}} - \tau_{n-1}^{\text{ep}}\right) : N_n}{H_{\text{s}} + G_{\text{e}}}. \tag{13}$$

Subsequently, τ^{ep} and β at the current time step are obtained by

$$\tau_n^{\text{ep}} = \tau^{\text{ep,trial}} - G_{\text{e}}\lambda N_n. \tag{14}$$

and

$$\beta_n = \beta_{n-1} + \frac{H_{\text{p}}}{H_{\text{s}}}\left(\tau_n^{\text{ep}} - \tau_{n-1}^{\text{ep}}\right). \tag{15}$$

3. Discussion and results

Measurement data from Lejon and Kari [10] was used for the parameter identification. 20 °C was chosen and the strain amplitudes are 0.0005, 0.0015 and 0.005. The frequency for the loading ranging from 200 to 900 Hz. More details for the measurement setup and method is referred to ref. [10]. Firstly, magnetic independent parameters include G_{e0}, $S_{\text{bounding0}}$, H_{p0}, a and b were obtained by the nonlinear least square method with "*lsqnonlin*" function in matlab® (MATLAB Release 2015b, The MathWorks, Inc., Natick, Massachusetts, United States). The results are $G_{\text{e0}} = 4.747 \times 10^6$ Pa, $S_{\text{bounding0}} = 1.528 \times 10^4$ Pa, $H_{\text{p0}} = 2.765 \times 10^7$ Pa, $a = 0.283$ and $b = 0.330 \times 10^6$ Ns^{a}m^{-2} with a relative error between measurement and simulation results 2.835%. After the first round of parameter identification, magnetic dependent parameters were identified under three kinds of magnetic field which are 0, 0.3 and 0.55T respectively. The results are $\kappa_1 = 0.418$, $\kappa_2 = 0$, $\kappa_3 = 0.387$ and $I_{\text{s}} = \left(0.394\text{T}/\mu_0\right)^2$ with μ_0 is the magnetic permeability in vacuum. The relative error between the measurement and simulation results is 3.501%. The comparison between the measurement and simulation result are shown from figure 2 to 4. It can be found that the match between the test and modelling result are quite good except the jump for the magnitude and loss factor in the frequency range from 600 to 700 Hz which is caused by the resonance of the test machine. Therefore, with the constitutive model developed in this paper, the amplitude, frequency and magnetic dependency of MS rubber can be predicted with a high accuracy.

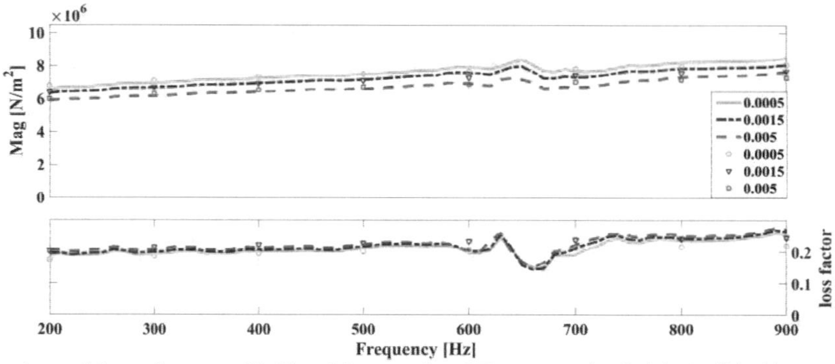

Figure 2. Magnitude and loss factor of MS rubber under 0T magnetic field. Solids lines and symbols are experimental and simulation result, respectively. 0.0005, 0.0015 and 0.005 are strain amplitudes.

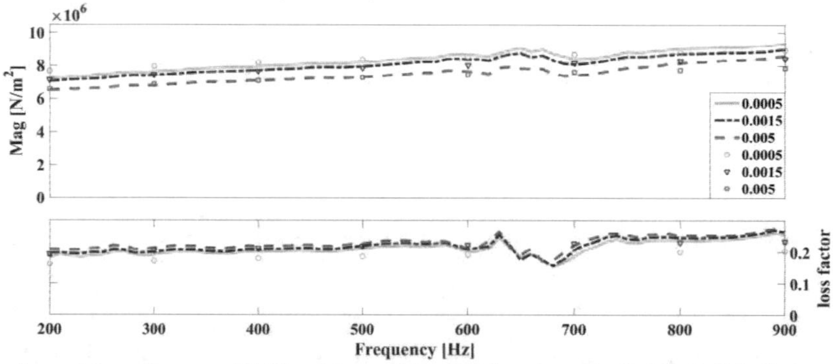

Figure 3. Magnitude and loss factor of MS rubber under 0.3T magnetic field. Solids lines and symbols are experimental and simulation result, respectively. 0.0005, 0.0015 and 0.005 are strain amplitudes.

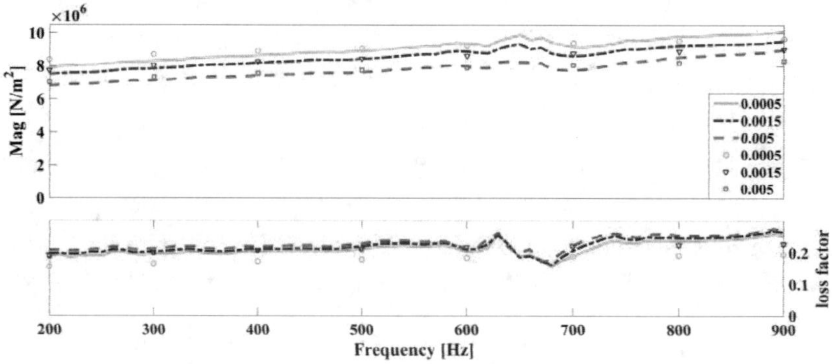

Figure 4. Magnitude and loss factor of MS rubber under 0.55T magnetic field. Solids lines and symbols are experimental and simulation result, respectively. 0.0005, 0.0015 and 0.005 are strain amplitudes.

4. Conclusion

A new constitutive model of isotropic MS rubber under shear strain was proposed in this paper. Fractional derivative element and a nonlinear kinematic elastoplastic model was used to represent the frequency and amplitude dependency of MS rubber, respectively. The magnetic dependency is reflected by a hyperbolic tangent function with considering of the magnetic saturation. After parameter identification, the conclusion that the amplitude, frequency and magnetic dependency of MS rubber can be depict by this proposed constitutive model can be drawn. The work in this paper fills the research deficiency of amplitude dependency in constitutive modelling of MS rubber and sets a foundation for the prediction of MS rubber-based vibration reduction devices and the possible application of those devices in noise and vibration control area.

IOP Conf. Series: Materials Science and Engineering **855** (2020) 012002 doi:10.1088/1757-899X/855/1/012002

5. Reference

[1] Deng H and Gong X 2007, *Journal of intelligent material systems and structures* **18** 1205-10
[2] Wang B and Kari L 2019, *Smart materials and structures* **28** 065026
[3] Jung H, Eem S, Jiang D and Koo J 2011, *Journal of intelligent material systems and structures* **22** 1439-50
[4] Sun S, Yang J, Du H, Zhang S, Yan T, Nakano M and Li W 2018, Journal of intelligent material systems and structures **29** 1777-89
[5] Yang J, Sun S, Tian T, Li W, Du H, Alici G and Nakano M 2016, *Mechanical systems and signal processing* **70-71** 811-20
[6] Jolly M R, Carlson J D and Muñoz B C 1996, *Smart materials and structures* **5** 607-14
[7] Kon Y, Jin K, Xu L and Zheng X 2017, *Smart materials and strucutres* **26** 115017
[8] Dorfmann A and Ogden R W 2004, *The quarterly journal of mechanics and applied mathematics* **57** 599-622
[9] Bustamante R 2010, *Acta mechanica* **210** 183-214
[10] Lejon J and Kari L 2013, *Journal of vibration and acoustics* **135** 064506
[11] Haupt P and Lion A 2002, *Acta mechanica* **159** 87-124
[12] Österlöf R, Wentzel H and Kari L 2016, *International journal of plasticity* **87** 1-14
[13] Mróz Z, Shrivastava H P and Dubey R N 1976, *Acta mechanica* **25** 51-61
[14] Dafalias Y F and Popov E P 1977, *Nuclear engineering and design* **41** 293-302

IOP Conf. Series: Materials Science and Engineering **855** (2020) 012003 doi:10.1088/1757-899X/855/1/012003

Functionalised Composite Materials for the Inflatable Modular System Accomplishment Used as Decoy for the Instruction and Training Maritime Applications

C Mihai[1] and A G Ene[1*]

[1]IT Research Department in Industrial Engineering, The Research-Development National Institute for Textile and Leather, 16 Lucretiu Patrascanu Street, Sector 3, 030508, Bucharest, Romania.

[*]alexandra.ene@incdtp.ro

Abstract. The inflatable modular systems are currently used in training and instruction specific to maritime applications. The main requirements imposed to this kind of systems are represented by: water and air impermeability; high resistance to the UV radiation action, to hydrolysis and to abrasion; resistance both to chemicals contained in seawater, and to the action of microorganisms, bacteria, mildews etc. For the accomplishment of the modular system was develop a composite material having the following physical and mechanical characteristics: mass: 173 $gsqm^{-1}$; breaking resistance: $min.72\ daN5cm^{-1}$ (for both systems); breaking elongation: min. 40% for weft and warp; air permeability: 0 cfm; resistance to water penetration: 1000 mm H_2O col. The floating target was designed considering the minimum requirements imposed to the system, regarding the maximum tension of the basic supports that must not exceed the allowed tension, for the wind speeds of 31 ms^{-1}, a sea agitation of 7 bf. The experimental programme performed in collaboration with the maritime authorities demonstrated that the inflatable modular system has the dimensional, gravimetric and functional characteristics that correspond to the requirements imposed by the final application.

1. Introduction

The implementation of modern techniques and training systems represents a desire of the personnel involved in maritime activities. Thus, in the naval field, one of the strategic objectives is the training and instruction of the personnel using modern, light equipment, which will ensure the increase of the technical-tactical accuracy.

Currently, in the ordinary applications, a series of floating target systems made of metal parts with heavy construction that requires additional human and financial efforts for maintenance are used. The most important inconvenience recorded for these types of systems are related to [1]:

- the destruction of one or more floating elements as a result of the targeting errors, that can lead to partial or total sinking of the target. In these situations, the costs associated with the operations of the target inflating, towing and repairing of these floating elements are extremely high;

- the usage of a complex anchoring systems consisting of anchors and special frames, resistant to mechanical shocks, precisely dimensioned so as to ensure that the target with the weight of the order of the tones is maintained at a fixed point;
- commitment of very high expenses for the transport and stationing of the special naves that ensure the towing and retraction of the targets up to and from the training area, in condition of maintaining a distance of at least 20 km from the target to the nave.

The innovative alternative development solution for these type of systems lies in the replacement of the metal materials with composite textiles so as the following technic and strategic requirements to be ensured: floatability under severe weather conditions; high corrosion resistance specific to the marine medium usage (e.g. microorganisms, bacteria, mildews, salt concentration etc.); high mechanical stability and resistance to environmental factors (e.g. waves between 1 and 2 m, speed of the wind of min.6 m/s, sea currents and periodical agitation); contrasting sea color to be observed from a great distance; constructive shape that allows a maximum viewing surface and optimal positioning at sea.

2. Material and methods

The floating target was designed considering the following minimum requirements imposed to the system:

- the maximum tension of the basic supports must not exceed the allowed tension, for the wind speeds of 31 ms^{-1}, and a sea agitation of 7 *bf;*
- the minimum strain in the system should be of stretching, although local deformations of the surfaces can be tolerated;
- the instability of the system on the wave, due to the wind load, or its detachment from the surface of the water, due to the waves should be avoided;
- the stresses developed in the base supports must be such that, added to the external forces, to satisfy the three requirements mentioned above.

The system dimensioning was accomplished based on the specific constructive requirements regarding [2]:

- the anchorage that should be made in three points (with an inert weight) and a safety factor of 1.5 in order enable the resistance loads that appear at the contact of the system surface with the water surface. It is important to consider that the anchoring system must develop maximum forces on the perimeter of the surface;
- the collapse time due to the shock, wind speed and intensity, excessive loads, should be less than two minutes, otherwise the local folding of the floating target can lead to strong bending movements. This phenomenon could determine an aerodynamic configuration that determine the breaking of the base material.

The main components of the system are represented by the: floating element, bellow, corner and reflector. The main component of the inflatable maritime system that ensure the its' floatability is the floating element of cylindrical form, made of a composite textile through sewn and sealed assemblies (figure 1). The bellows represent the outer envelope of the floating element, with the role of protection against friction, scratches, stabs. It takes over the mechanical stresses, mainly due to the inflation pressure of the floating component (figure 2). Optimum colors for maximum visual imprint are red or orange.

IOP Conf. Series: Materials Science and Engineering **855** (2020) 012003 doi:10.1088/1757-899X/855/1/012003

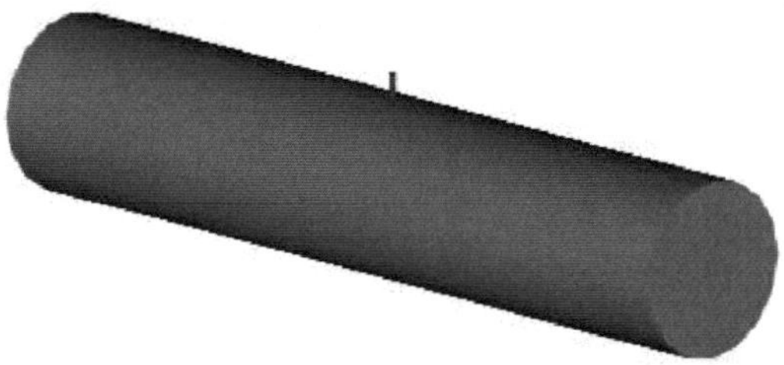

Figure 1. The floating element of the inflatable maritime system.

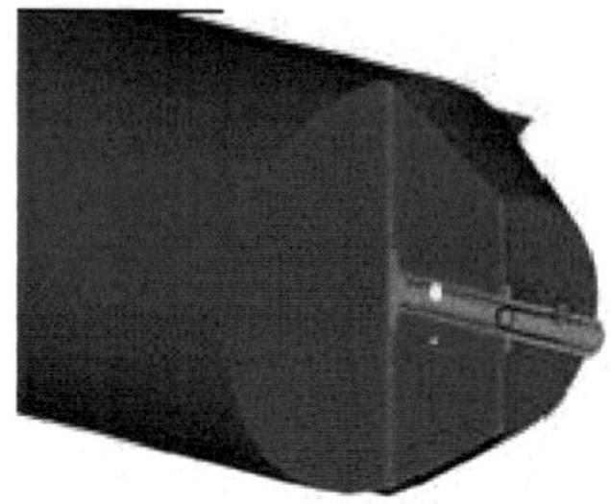

Figure 2. The bellows of the floating system.

The corner (figure 3) is the connecting piece that enables the assembly of several bellows in a homogeneous construction (figure 4). It is obtained by processing of the polypropylene under pressure in injection molds.

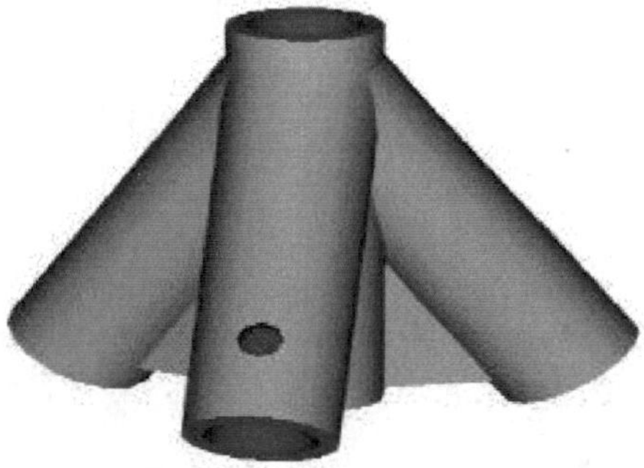

Figure 3. The corner.

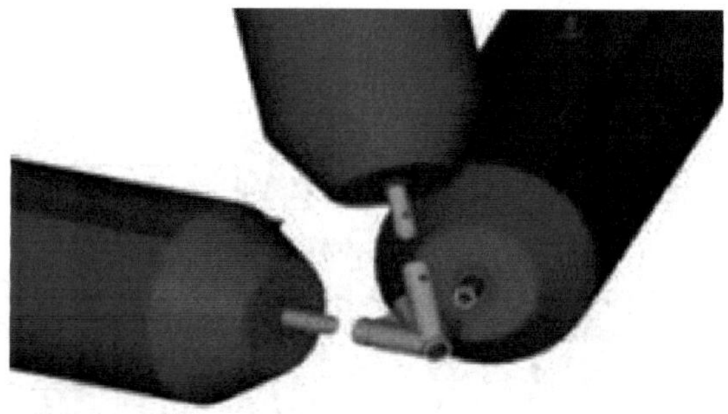

Figure 4. Assemble of several bellows

The reflector has the role of amplifying the visual imprint. It is made of red or orange textile panels.
For the study of the forces, buoyancy and stability of the structure, the following working hypotheses were considered:

- the bellows have the length L = 1 m, and the diameter D = 0.15 m,
- the bellows are completely submerged in water,
- Archimedean force is uniformly distributed along the length of the cylinder,

Using the simple leaning bar model [3] with the uniformly distributed load (figure 5 and figure 6) the following relations was obtained:

$F_A = 9{,}8 \cdot V \cdot \rho$, where: V = volume of the cylinder (m^3); ρ = the density of the liquid kgm^{-3}.

$p = F_A \cdot L^{-1}$, where: p = load uniform distributed ($N \cdot m^{-1}$); F_A = Archimedean force (N); L = the length of the cylinder (m)

$M_{max} = p \cdot L^2 \cdot 0{,}125$, where: M_{max} = maximum bending moment (m^3)

Using these formulas, it can be obtained the following: $F = 173{,}18\ N, p = 173, 18\ N \cdot m^{-1}$, $M_{max} = 21{,}65\ Nm$.

It should be emphasized that the bending moment introduces uniform distributed loads in the cylinder shell, stretching load in the upper part, respectively compression in the lower part, because it appears in the middle section of the cylinder.
Since the bellows cannot take on compression loads, it must be pre-tensioned in a free state (not submerged) at a load at least equal to δ_{max} [4].
This aspect determines the doubling of the unitary effort δ_{max} in the submerged state, respectively: $\delta_{max} = 4920\ N \cdot m^{-1}$.

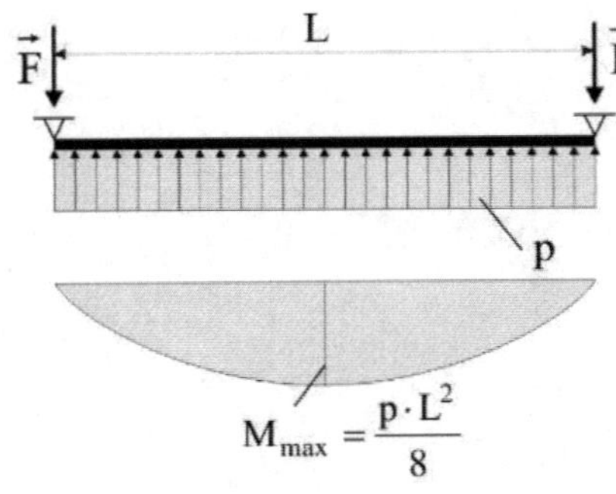

Figure 5. Forces distributed in simple leaning bar model.

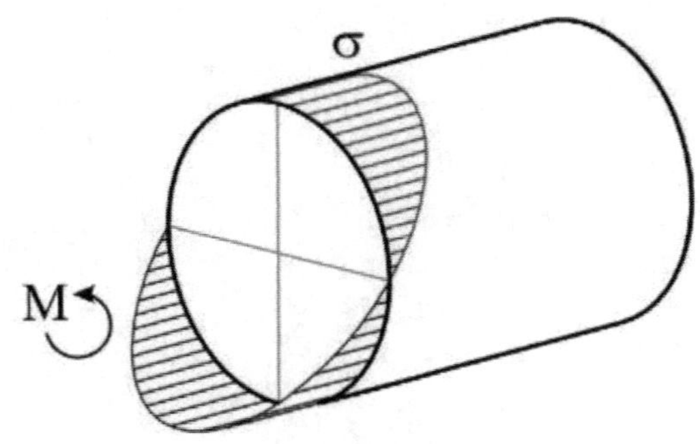

Figure 6. The bending moment.

The resulting values for the volume of dislocated water (respectively Archimedean force) at the inflatable modular system submersed at different depths are presented in table 1 and figure 7.
The values obtained for the moment of return to the vertical position (figure 8) of a submerged inflatable modular system at depths in the range [0; 7] cm is presented in table 2.

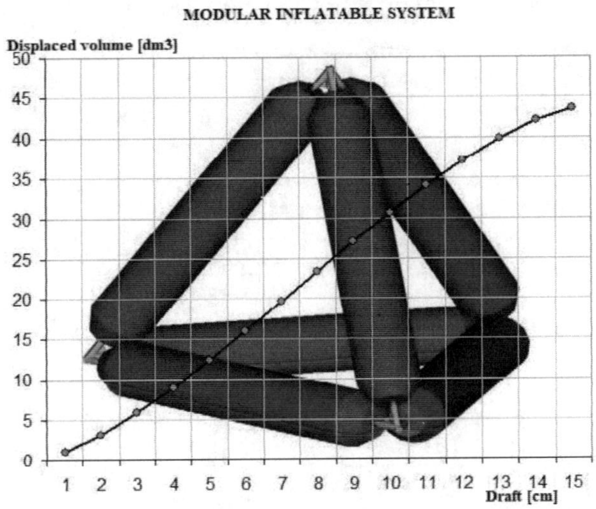

Figure 7. Floatability of the inflated modular system.

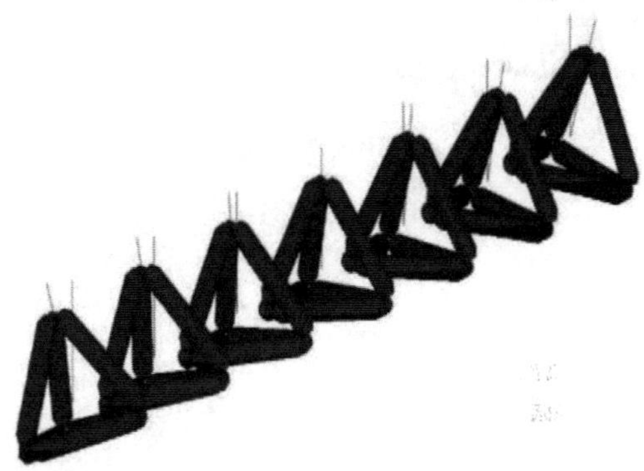

Figure 8. Sequential representation of the inflatable modular system returns in vertical position.

Table 1. The volume of the dislocated water at the submersing of the modular system

Draft (m)	Displaced volume (m^3)	Archimedean forces (daN)
1	0.0011	1.078
2	0.0032	3.136
3	0.0059	5.782
4	0.009	8.820
5	0.0124	12.152
6	0.0159	15.582
7	0.0196	19.208
8	0.0233	22.834
9	0.027	26.460
10	0.0306	29.988

IOP Conf. Series: Materials Science and Engineering **855** (2020) 012003 doi:10.1088/1757-899X/855/1/012003

11	0.034	33.320
12	0.0371	36.358
13	0.0398	39.004
14	0.0421	41.258

The inflatable modular system can constitute a cell of a matrix (figure 9). The assembly mode is made simple and fast through of a "fork" (figure 10).

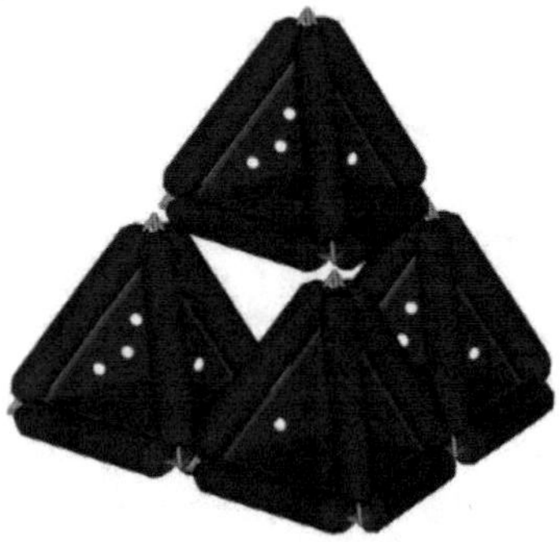

Figure 9. Cell of the matrix

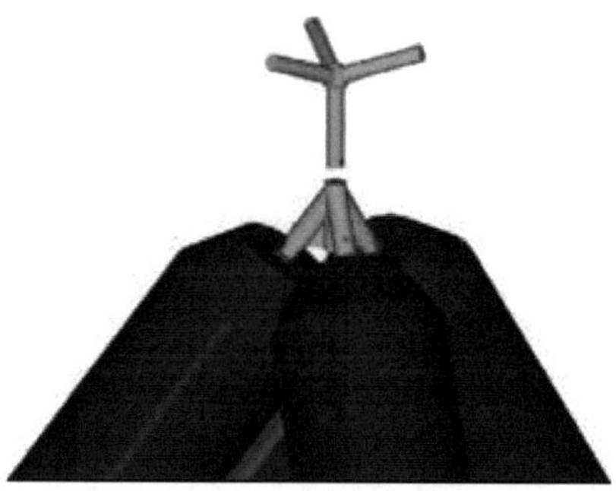

Figure 10. Assembly system.

Verification and testing of the inflatable modular system under the open sea were carried out under the following conditions: wind speed: $6\ m \cdot s^{-1}$; air temperature: $25°\ C$; water temperature: $25°\ C$. It was evidenced that the floatability of the system was 100%. In addition, the stability and visibility of the naval floating target throughout the application were kept constant.

Table 2. The moment of the inflatable modular system in vertical position

		Displaced volume (m^3)	Y (m)	Return moment (Nm)
Draft = 0-1 cm	-15°	0.0070 - 0.0087	0.212 – 0.206	14.84 - 17.90
	-10°	0.0042 - 0.0058	0.234 - 0.225	9.84 - 13.03
	-5°	0.0016 - 0.0030	0.255 - 0.235	4.08 - 7.04
	0°	0.0000 - 0.0011	0.000	0.00
	5°	0.0018 - 0.0031	(-0.328) – (-0.296)	(-5.90) – (-9.18)
	10°	0.0047 - 0.0063	(-0.311) - (-0.294)	(-14.61) - (-18.52)
	15°	0.0076 - 0.0091	(-0.288) - (-0.276)	(-21.86) - (-25.11)
Draft = 2-3 cm	-15°	0.0105 - 0.0125	0.199 - 0.191	20.87 - 23.93
	-10°	0.0077 - 0.0098	0.214 - 0.203	16.48 - 19.85
	-5°	0.0048 - 0.0070	0.212 - 0.187	10.16 - 13.06
	0°	0.0032 - 0.0059	0.000	0.00
	5°	0.0048 - 0.0068	(-0.264) - (-0.230)	(-12.67) - (-15.63)
	10°	0.0078 - 0.0096	(-0.277) - (-0.259)	(-21.58) - (-24.84)
	15°	0.0106 - 0.0123	(-0.264) - (-0.252)	(-28.02) - (-31.02)
Draft = 4-5 cm	-15°	0.0146 - 0.0166	0.183 - 0.175	26.78 - 29.00
	-10°	0.0121 - 0.0145	0.190 - 0.177	23.01 - 25.68
	-5°	0.0097 - 0.0128	0.160 - 0.133	15.49 - 17.04
	0°	0.0090 - 0.0124	0.000 -	0.00
	5°	0.0095 - 0.0126	(-0.183) - (-0.146)	(-17.39) - (-18.36)
	10°	0.0116 - 0.0137	(-0.241) - (-0.224)	(-27.92) - (-30.70)
	15°	0.0140 - 0.0157	(-0.240) - (-0.226)	(-33.54) - (-35.51)
Draft = 6-7 cm	-15°	0.0187 - 0.0070	0.165 - 0.212	30.93 - 14.84
	-10°	0.0171 - 0.0042	0.163 - 0.234	27.89 - 9.84
	-5°	0.0161 - 0.0016	0.111 - 0.255	17.82 - 4.08
	0°	0.0159 - 0.0000	0.000	0.00

5°	0.0159 - 0.0018	(-0.117) - (-0.328)	(-18.59) - (-5.90)
10°	0.0161 - 0.0047	(-0.203) - (-0.311)	(-32.60) - (-14.61)
15°	0.0175 - 0.0076	(-0.212) - (-0.288)	(-37.17) - (-21.86)

3. Results and discussion

The experimental programme performed in collaboration with the maritime authorities demonstrated that the inflatable modular system has the dimensional, gravimetric and functional characteristics that correspond to the requirements imposed by the final application, respectively:

- in dynamic conditions (wave - wind) under the real weather conditions the proper functionality of the system was proved;
- the resistance to water penetration was adequate and the system was maintained on the initial floating line. Thus, was demonstrated that the outer envelope remained sealed during the test and the water did not penetrate inside;
- the tetrahedron designed constructive form, is in accordance with the technical-tactical requirements, and the orange colour is contrasting with that of the sea, enabling very good visibility;
- the selected materials, the architecture of the modular system and the established constructive dimensions, equip the inflatable modular system with a valuable set of technical characteristics related to stability, floatability, visibility etc.

4. Conclusion

The inflatable modular system demonstrated high resistance during a considerable number of training applications (more than 10.000), the following characteristics being maintain accordingly:

- stability on the wave, visibility and floatability under hydrometeorological testing conditions;
- the scattering ellipse of the training elements, determined by the naval action system, was in compliance with the technical and tactical requirements specific to the instruction and training applications, which showed that the unconventional accomplished system, was buoyant, visible and stable within the determined and predetermined perimeter.

The inflatable modular system is removable, and can be stored in small spaces, without special arrangements, on nave board.

5. References

[1] Sánchez-Gálvez V, Cendón D, Sancho R, Gálvez F 2019 Performance of composite materials under high-speed impact in extreme environments *Multiscale and Multidisciplinary Modeling, Experiments and Design* 2 97-104

[2] Sánchez-Gálvez V, Gálvez F, Sancho R, Cendón D 2017 A new analytical model to simulate high-speed impact onto composite materials targets *Int J Impact Engng* **108** 322–333

[3] Wiedenman N, Dharan C K H 2006 Ballistic penetration of compressively loaded composite plates *Journal of Composite Materials* **40** 12 1041-61

[4] Yuan Z, Crouch R, Wollschlager J, Fish, J 2107 Assessment of multiscale designer for fatigue life prediction of advanced composite aircraft structures *J. Compos. Mater.* **51** 2131–41

Acknowledgment

This scientific paper is funded by the Ministry of Research and Innovation within Program 1 - Development of the national RD system, Subprogram 1.2 - Institutional Performance - RDI excellence funding projects, Contract no. 6PFE/2018.

IOP Conf. Series: Materials Science and Engineering **855** (2020) 012004 doi:10.1088/1757-899X/855/1/012004

Analysis of vibration response from high damping rubber bearing and lead rubber bearing for using in high rise building foundations subjected to earthquake

N Alifani [1, a] and W Nirbito [1, b*]

[1] Department of Mechanical Engineering, Faculty of Engineering, Universitas Indonesia, Kampus Baru UI Depok, Depok 16424, Indonesia

[a] nabila.alifani@yahoo.com; [b] bito@eng.ui.ac.id

Abstract. Vibration due to earthquakes that occur in a high rise building like hospital building can disrupt the comfort, as well as operation of the hospital, and can cause casualties because usually, the building response to the earthquake is of a cantilever movement. The solution from a mechanical vibration point of view is to change the response of buildings to earthquakes so that hospital buildings are not destroyed by earthquakes by controlling the response of vibrations that produce lateral movement responses. One technology that is used to increase the resilience of buildings that subjected to earthquake, as well as control vibration response is the application of rubber bearing base isolation on the building foundations. High damping and lead rubber bearings base isolation are used on high rise building foundations which are designed to give lateral motion due to the base isolation construction. The results of computer simulation using ANSYS® software have shown that the rubber bearing base isolation gave vibration response in lateral motion.

1. Introduction

The impact of a strong earthquake response on a hospital building which is full of patients, medical personnel and medical care facilities, as well as many modern sophisticated types of equipment, is very dangerous. This can happen since the hospital building is rigid so the vibration response to the earthquake will be in cantilever motion. The solution to controlling such a vibration response is applying the principle of mechanical vibration that will respond in a lateral movement. One of the earthquake-resistant structural building is by using rubber bearing base isolation on the foundation. The types of high damping rubber bearing and lead rubber bearing base isolation. However, it is not yet known how the real vibration response of rubber bearing base isolation types will be when an earthquake occurs. So, therefore, virtual reality simulation using ANSYS® software was conducted to determine the vibration response of the high damping and lead rubber bearing base isolation due to the earthquake.

2. Analysis model

2.1. Specifications of high damping rubber bearing

The base isolation type high damping rubber bearing used in the hospital building studied is made of natural rubber sheets affixed to steel plates alternately which has no lead core in the center with the damping ratio (ζ) on the base isolation type high damping rubber bearing is 0.240 [1] and has the following specifications presented in Table 1. High damping rubber bearing is modeled as in Figure1. With the red layer being the rubbers and the black layers being steel plates. [2]

Table 1. Design characteristics of high damping rubber bearing [1]

Component			Value
Outer diameter	d_0	mm	800
Number of rubber layers	n	-	37
Thickness of rubber layers	t_r	mm	5.4
Height	H	mm	422.2
Shear modulus	G_{eq}	N/mm^2	0.620
Spring constant	K	10^3 kN/m	1.56

2.2. Specifications of lead rubber bearing

The base isolation type lead damping rubber bearing used in the hospital building studied is made of natural rubber sheets affixed to steel plates alternately which lead core in the center with the damping ratio (ζ) on the base isolation type lead damping rubber bearing is 0.300 [1] and has the following specifications presented in Table 2. Lead rubber bearing is modeled as in Figure 2. With the red layer being the rubbers, the black layers being steel plates and the green being lead. [3]

Table 2. Design characteristics of lead damping rubber bearing [1]

Component			Value
Outer diameter	d_0	mm	800
Number of rubber layers	n	-	37
Thickness of rubber layers	t_r	mm	5.4
Height	H	mm	422.2
Outer diameter of lead plug	d_l	mm	180
Shear modulus	G_{eq}	N/mm^2	0.385
Spring constant	K	10^3 kN/m	2.01

Figure 1. High damping rubber bearing modeled in autodesk inventor

Figure 2. Lead rubber bearing modeled in autodesk inventor

3. Simulation parameters

Base isolation used in hospital building totaled 161, but in this study, only one base isolation was simulated and analyzed, so the mass of hospital building used in calculations and simulation was the mass of hospital building for one base isolation. It is calculated that the mass of the hospital building is

259,989.666 kg with the maximum earthquake acceleration based on Figure 3 is 0.554 g during the earthquake period (T) of 0.57 seconds, so the earthquake frequency (f_p) is 1.75 Hz.

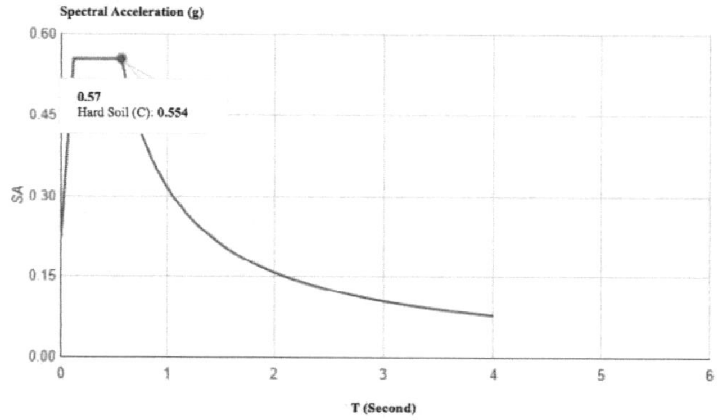

Figure 3. The earthquake excitation (http://puskim.pu.go.id/) [4]

To determine the earthquake excitation force using the following the Eq. (1).

$$F(t) = Fcos\omega_t \tag{1}$$

The desired earthquake excitation is at the time of the maximum earthquake, so the value of $cos\omega_t$ is 1. To determine the maximum earthquake excitation using the following Eq. (2). The result of the maximum earthquake excitation force (F(t)) is 1,412,976.237 N.

$$F = m \cdot a \tag{2}$$

To analyze the vibration response of the base isolation type high damping rubber bearing and lead rubber bearing are with a frequency ratio using the following Eq. (3). The result of the frequency ratio at the time of earthquake frequency (f_p) is 1.75 Hz at base isolation type of high damping rubber bearing is 4.6 while the type of lead rubber bearing is 4.16.

$$\beta = \frac{\omega_f}{\omega_n} = \frac{f_p}{f_n} \tag{3}$$

Based on the magnification factor vs frequency ratio graph in Figure 4, the vibration response on the base isolation type of high damping rubber bearing and lead rubber bearing that occurs when the earthquake frequency (f_p) is 1.75 Hz is very small, which means the vibration amplitude is very small. The magnification factor value produced is very small due to the small value ω_n. The cause of the small value ω_n is due to base isolation that is not rigid.

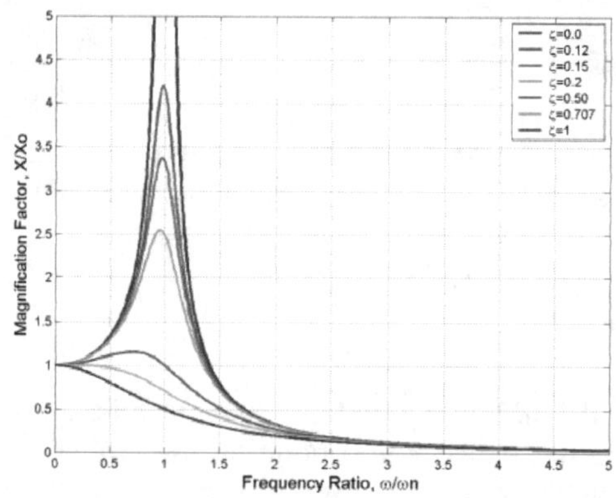

Figure 4. Magnification factor vs frequency ratio [5]

4. Simulation preparations

To find out the vibration response of the base isolation type high damping rubber bearing and lead rubber bearing, simplified modeling using FEA in ANSYS®. In this study, two types of simulation are ran using the harmonic response analysis tools. The first is high damping rubber bearing simulations and the second is lead rubber bearing simulations. Simulations using models and specifications of high damping rubber bearings and lead rubber bearings are presented in Table 1 and Table 2 and the calculated harmonic response parameters inputted.
For a clearer visualization, a base is added at the top and bottom of the base isolation model. The top base is assumed as the building and the bottom base as the ground. A distributed mass of 259,989.666 kg is applied vertically to the top base and an excitation force of 1,412,976.237 N is applied horizontally at the bottom base. The pre-simulated model is shown in Figure 5.

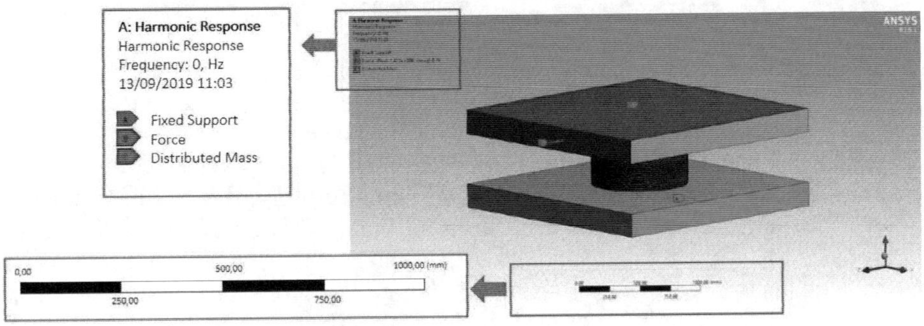

Figure 5. The pre simulated model for the ANSYS simulations

5. Result and discussion

The simulation is done using FEA in ANSYS® software using the harmonic response analysis tools. The base isolation type high damping rubber bearing is modeled with 37 layers of rubber that affixed to steel plates alternately. The vibration response that occurs in the results of the simulation base isolation type of high damping rubber bearing in Figure 6 is the lateral movement. The lateral movement response that occurs in high damping rubber bearing in the face of earthquake exposure allows the hospital building to withstand the seismic impact which has the potential to destroy [6]. The results obtained from the simulation of the vibration response of high damping rubber bearing are displacement in the horizontal direction is 51.08 mm in Figure 6 and the vertical direction is 18.38 mm in Figure 7.

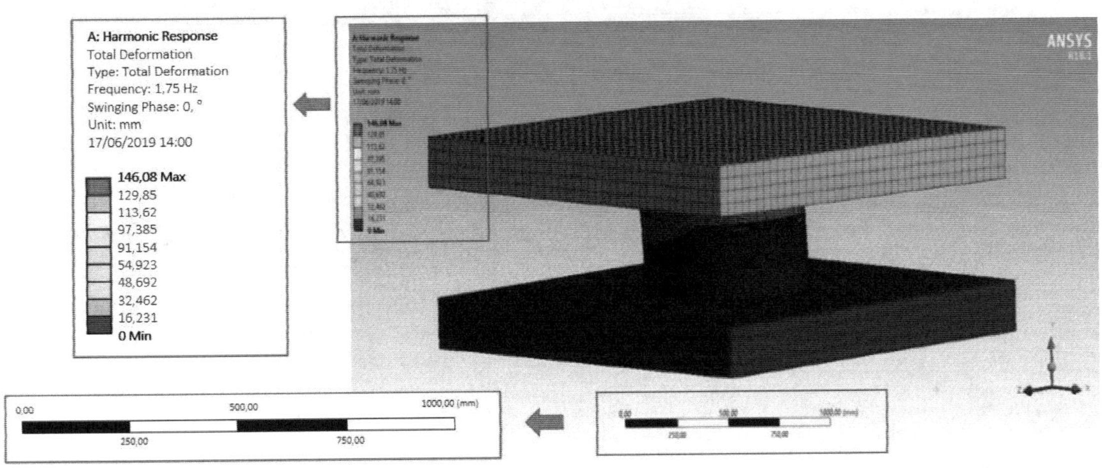

Figure 6. High damping rubber bearing simulation results at 1.75 Hz frequency in the horizontal direction.

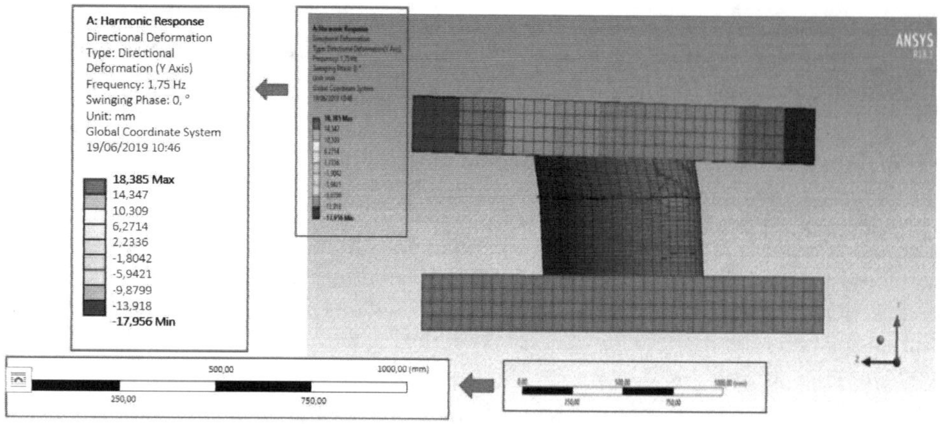

Figure 7. High damping rubber bearing simulation results at 1.75 Hz frequency in the vertical direction.

The base isolation type of lead rubber bearing is modeled with 37 layers of rubber that are affixed to the steel plates alternately and there is lead inside. The vibration response that occurs in the results of the simulation base isolation type of lead rubber bearing in Figure 8 is the response that approaches the cantilever movement. The response that approaches the cantilever movement that occurs can potentially damage the structure and contents of the hospital building. A response that approaches the cantilever movement can occur due to the addition of leads which makes the base isolation rigid. However, if seen from the simulation results of the vibration response of the base isolation type lead rubber bearing response that approaches the cantilever movement there is a relative motion between the layers of rubber and steel plate which allows a slight lateral movement, so that the response that approaches the cantilever movement on the base isolation type lead rubber bearing lead does not cause severe damage. The results obtained from the simulation of vibration responses of base isolation type lead rubber bearing are displacement in the horizontal direction is 33.27 mm in Figure 8 and the vertical direction is 30.57 mm in Figure 9.

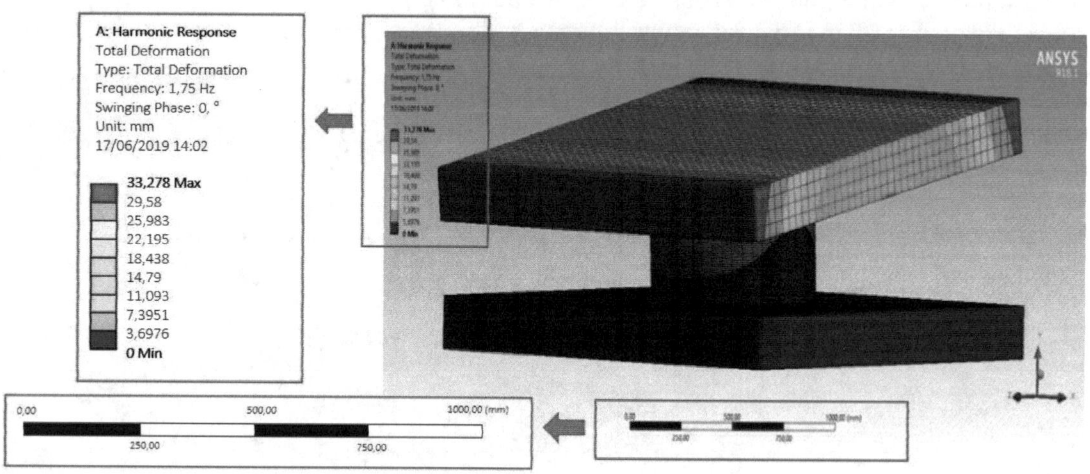

Figure 8. Lead rubber bearing simulation results at 1.75 Hz frequency in the horizontal direction.

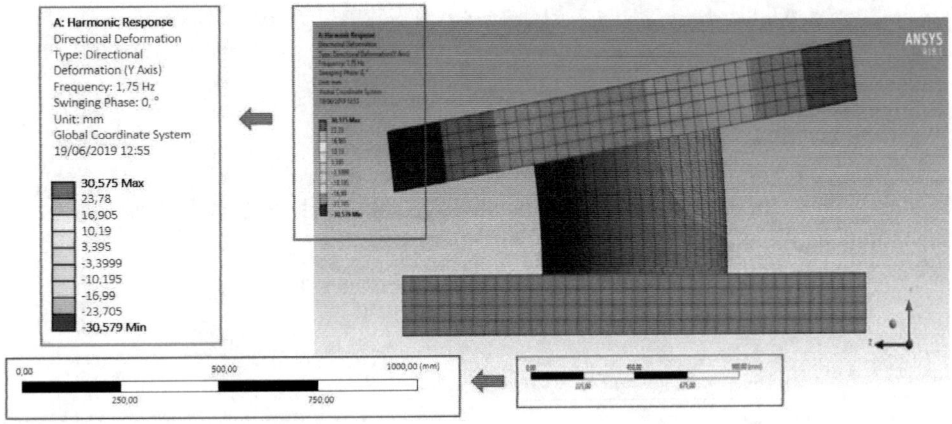

Figure 9. Lead rubber bearing simulation results at 1.75 Hz frequency in the vertical direction.

The results of the vibration response simulation of the base isolation type high damping rubber bearing produce lateral movement response and the base isolation type lead rubber bearing produces a response that approaches the cantilever movement. Base isolation type high damping rubber bearing is better than base isolation type lead rubber bearing in the face of earthquake exposure [7]. However, if the use of base isolation type lead rubber bearing combined with base isolation type high damping rubber bearing will make the base isolation type lead rubber bearings function as a displacement barrier, which means the displacement that occurs is smaller than the displacement on the base isolation type high damping rubber bearing.

6. Summary

By using base isolation type high damping rubber bearing, the vibration response during the earthquake period (T) 0.57 seconds is merely lateral movement. Displacement in the horizontal direction is 51.08 mm and in the vertical direction is 18.38 mm which is insignificant compared to horizontal displacement. Meanwhile, by using base isolation type lead rubber bearing, the vibration response during the earthquake period (T) 0.57 seconds is relatively still lateral movement since the total displacement is smaller. Displacement in the horizontal direction is 33.27 mm and the vertical direction is 30.57 mm. The base isolation is stiffer that makes the lateral movement shorter due to the addition of lead which makes the base isolation more solid.

If the use of base isolation is combined between types of high damping rubber bearing and lead rubber bearing, it is recommended that the lead rubber bearings are located in the outer circumference so it will function as a displacement barrier. Therefore total displacement is smaller than the displacement of all high damping rubber bearings only usage.

7. References

[1] PT. CAKRA MANGGILINGAN JAYA. 2018. *PLAN OF BASE ISOLATION DEVISES*. Jakarta

[2] DOSHIN RUBBER ENGINEERING-GO BEYOND LIMITS MAKE DREAMS REALIT. 2019. *Base Isolation High Damping Rubber Bearing*. Retrieved March 15, 2019, from Doshin Rubber Web site: https://doshinrubber.com

[3] DOSHIN RUBBER ENGINEERING-GO BEYOND LIMITS MAKE DREAMS REALIT. 2019. *Base Isolation Lead Rubber*. Retrieved March 15, 2019, from Doshin Rubber Web site : https://doshinrubber.com

[4] Pusat Penelitian Dan Pengembangan Pemukiman. 2019. *Desain Spektra Indonesia*. Retrieved March 10, 2019, from Pusat Penelitian Dan Pengembangan Web site: http://puskim.pu.go.id

[5] Rao, S. (2011). *Mechanical Vibrations* (5th ed.).

[6] Makoto O., Tomoshi M., Masayuki K., Takuzo Y., Masashi Y. & Naohiro N. 2015. *Finite Element Analysis of Laminated Rubber Bearing of Building Frame under Seismic Excitation*.

[7] Manasa M. S., Alice Mathai. 2017. *Performance of Lead Rubber Bearing as a Base Isolator International Journal of Science Technology & Engineering*, Volume 3, Issue 11.

Acknowledgments
This research was supported and funding by DRPM UI with grant PITTA No: NKB-0795/UN2.R3.1/HKP.05.00/2019. We thank to Universitas Indonesia Teaching Hospital and Henki Ashadi, PhD who provided insight that greatly assisted the research, although they may not agree with all of the interpretations or conclusions of this paper.

IOP Conf. Series: Materials Science and Engineering **855** (2020) 012005 doi:10.1088/1757-899X/855/1/012005

Optimization of On-Site Xylanase Production from Aspergillus niger via Central Composite Design (CCD)

M Tarrsini[1], Q H Ng[1], Y P Teoh[2, *], B Kunasundari[1, **], W U Ang[1], S H Shuit[3], Z X Ooi[4] and P Y Hoo[1]

[1]Department of Chemical Engineering Technology, Faculty of Engineering Technology, Universiti Malaysia Perlis, Kampus UniCITI Alam, Sungai Chuchuh, Padang Besar, 02100, Perlis, Malaysia.
[2]Department of Petrochemical Engineering, Faculty of Engineering and Green Technology, Universiti Tunku Abdul Rahman, Jalan Universiti, Kampar, Perak, Malaysia.
[3]Department of Chemical Engineering, Lee Kong Chian Faculty of Engineering and Science, Universiti Tunku Abdul Rahman, Sungai Long Campus, Jalan Sungai Long, Bandar Sungai Long, Cheras, 43000, Kajang, Selangor, Malaysia.
[4]Department of Chemical Science, Faculty of Science, Universiti Tunku Abdul Rahman, Jalan Universiti, Bandar Barat, 31900, Kampar, Perak, Malaysia.

Email: teohyp@utar.edu.my*; kunasundari@unimap.edu.my**

Abstract. Xylanases have stimulated considerable interest due to their potential application in several industries, especially in the bioethanol sector. Since the vitality of this enzyme is undeniable, this research is focused on optimization of on-site xylanase production from *Aspergillus niger* (*A. niger*). This initiative could reduce the dependence of commercial xylanase. Central Composite Design (CCD) was implemented in the process of xylanase production optimization. Incubation temperature and medium pH were two parameters selected to statistically optimized using Response Surface Methodology (RSM) in order to improve the xylanase production. From the data analyzed by Design of Experiment (DoE), maximal xylanase production was predicted to produce under condition of 32.67℃ and pH 4.56 with desirability of 0.936. A validation test with triplicate was done to verify the predicted result. The maximum enzyme activity of 0.5638 U/mL was obtained from the validation test.

1. Introduction

Xylanase is the name given to a classification of enzyme which degrade the second most abundant linear polysaccharide, beta-1,4-xylan into xylose, as a consequence of breaking down hemicellulose, one of the primary components of plant cell walls [1]. Xylanase is critical for the efficient use of plant biomass to produce fuels, energy and other chemical compounds. At present scenario, the secondary bioethanol industry is known to be dependent on xylanase for the biofuel production [2]. To be precise, secondary bioethanol production is based on lignocellulosic biomass with the polysaccharide composition of cellulose and hemicellulose which could be converted into biofuel. Thus, this

IOP Conf. Series: Materials Science and Engineering **855** (2020) 012005 doi:10.1088/1757-899X/855/1/012005

conversion process, biomass requires the aid of cellulose to co-exist with xylanase and cellulase to be breakdown into reducing sugar and consecutively fermented into bioethanol [3].

Yet, the utilization of commercial hydrolases could be a major factor in the elevation of the project's overall economics. With this, the credibility of the process in terms of being a cost-effective process will be in doubt [4]. Hence, on-site xylanase production was taken in consideration to reduce the expenditure on enzymes. Filamentous fungi are widely known to be the essential industrial sources of xylanase [5]. Among these microbial sources, *A. niger* is chosen to be used in this study because it has a specifically interesting recognition. This filamentous fungus secretes these enzymes into medium and their xylanase levels are much higher than those observed in yeasts or bacteria [6].

Generally, production of high extracellular enzymes with excellent industrial quantities are dependent on environmental conditions such as pH, temperature, initial moisture, inoculum sizes, incubation period and the nutritive aspects of the growth medium [7]. In other words, optimization of process parameters is important to enhance the activity and the volumetric productivity of general metabolites which include enzymes such as xylanase [8]. With that, the objective of this study is to optimize selected process parameters for maximal xylanase production by *A. niger* using the CCD of response surface statistical optimization approach.

2. Material and Method

2.1. Materials

All the chemicals used in this research work were analytical grade. These chemicals were purchased from Sigma-Aldrich, St. Louis, MO USA except otherwise stated.

2.2. Inoculum Development

A 10 mL of 1% (v/v) Tween-80-sterile distilled water was added to a 5 days old pure *A. niger* culture on agar plate and the agar surface was rubbed with a sterile L-shaped spreader to obtain spore suspension. The suspension was sieved through Whatman no.1 paper to remove mycelia. After appropriate dilutions, inoculum sizes (2 x 10^6 spores/mL) were determined by direct microscopic observation using hemocytometer.

2.3. Submerged Fermentation (SmF)

Xylanase biosynthesis were performed in a 250 mL Erlenmeyer flask using SmF in a production medium containing 100 mL of Mandels and Sternburg's basal medium and 5.0 g/L of xylan as carbon source. Composition of the growth medium includes: 2.0 g/L NH_4NO_3, 2.0 g/L K_2HPO_4, 1.0 g/L $MgSO_4.7H_2O$, 0.3 g/L $CaCl_2$, and 0.012 g/L trace elements: 5.0 mg/L $FeSO_4.7H_2O$, 1.6 mg/L $MnSO_4.4H_2O$, 3.45 mg/L $ZnSO_4.7H_2O$ and 2.0 mg/L $CoCl_2.6H_2O$ and 0.1% (v/v) Tween 80. Aliquot of 1 mL of standardized *A. niger* inoculum (2 x 10^6 spores/mL) was aseptically inoculated into the medium and fermentation was allowed to take place at different selected incubation temperature at 150 rpm for 5 days in an incubator shaker (Jeiotech, Korea). Crude enzyme was extracted with 0.05 M citrate buffer pH 4.8 (50 mL). After the addition of the buffer, the whole flask was kept in a rotary incubator at 150 rpm (32°C) for 15 min. The content was filtered through Whatman no.1 filter paper to extract the filtrate which was then centrifuged at 10,000 rpm for 15 min at 4°C. The supernatant was collected as the crude enzyme for xylanase activity analysis [9].

2.4. Optimization Process of Xylanase Production

Response surface methodology was used for the optimization of conditions for xylanase production by *A. niger* using Central Composite Design (CCD) of the Design expert DX 11.0.4.0. A total of 11 runs including 4 factorial points, 4 axial points, and 3 center points in triplicates were independently done

IOP Conf. Series: Materials Science and Engineering **855** (2020) 012005 doi:10.1088/1757-899X/855/1/012005

to verify the effects of variables on xylanase production. The variables studied in this model was incubation temperature (X_1) and medium pH (X_2). Each parameter was divided into three levels: low (-1), medium (0) and high (+1). Optimum incubation temperature and medium pH derived from CCD optimization were used as functions to generate model for enzyme activity (U/mL). The model was analysed based on ANOVA and regression analyses.

3. Result and Discussion

The overall design matrix of experiments on optimization of xylanase production from *A. niger* coupled with the experimental enzyme activity via CCD is shown in Table 1. pH and temperature of the cultivation medium were selected to be optimized value were the most significant variables for enzyme production by *Aspergillus* species [10]. Based on collected data, the enzyme activity was found varying in the range of 0.5237 U/mL to 0.3368 U/mL. The experimental run with highest enzyme activity (0.5237 U/mL) was achieved with run #1 (32.50°C, pH 5.00). In contrast, the lowest enzyme activity (0.3368 U/mL) was measured at run #7 (38.86°C, pH 5.00).

Table 1: Experimental design matrix of xylanase production from *A. niger* using CCD.

Std	Run	Factors		Response	
		Incubation temperature (°C)	Medium pH	Enzyme activity (U/mL)	Standard deviation
		(X_1)	(X_2)		
9	1	32.50	5.00	0.5237	0.0008
8	2	32.50	7.83	0.3990	0.0046
11	3	32.50	5.00	0.5078	0.0010
7	4	32.50	2.17	0.4336	0.0046
10	5	32.50	5.00	0.4974	0.0025
4	6	37.00	7.00	0.4075	0.0015
6	7	38.86	5.00	0.3368	0.0039
1	8	28.00	3.00	0.4061	0.0051
3	9	28.00	7.00	0.3543	0.0014
5	10	26.14	5.00	0.3787	0.0017
2	11	37.00	3.00	0.4557	0.0018

* The shaded rows indicated the maximum (run #1) and minimum (run #7) enzyme activity.

The analysis of variance (ANOVA) was primarily used to identify the mathematical model that best suit for the condition. The model was selected based on the degree of significance. The significance of each coefficient was determined by whether it had p-value which less than 0.05. The F-value is the measurement of variation of the data about the mean. The high F-value and very low probability indicate that the model is in good prediction of the experimental results [11].

Table 2 tabulates the analysis of variance table for response quadratic model. Based on Table 2, it can be concluded that quadratic model was the best model to fit the response as the F-value was observed to be 7.36 implied that the model was significant. There was only a 2.36% chance that a "Model F-Value" could occur due to noise. The "Lack of Fit F-value" of 8.24 implied that the model is not significant relative to the pure error (more than 0.05). However, there is a 11.02% of chance that a "Lack of Fit F-value" could occur due to noise. Insignificant of Lack of Fit indicated that the model was significant and fit. Besides, the accuracy of the fit of the model was checked by multiple correlation coefficients (R^2). The closest the correlation R^2 to value of 1.00, the higher the quality of the model. From the results of this study, the value of R^2 was found to be 0.8803 indicated that there was a high correlation of experimental value towards predicted value.

Table 2: Analysis of variance table for response surface quadratic model.

Source	Sum of Squares	df	Mean Square	*F* Value	*p*-value Prof > *F*	Remark
Model	0.0344	5	0.0069	7.36	0.0236	significant
Residual	0.0047	5	0.0009			
Lack of Fit	0.0043	3	0.0014	8.24	0.1102	not significant
Pure Error	0.0004	2	0.0002			
Cor Total	0.0391	10				

*Cor Total= Corrected Total; df= Degree of freedom

Furthermore, second order equations were developed from the analysis of variance. The polynomial models from experimental results for enzyme activity were derived as shown in Equation 1 (actual factors terms) to assist future research works related to production of xylanase from *A. niger*.

Enzyme activity=
-3.44396 + 0.229322 (Incubation temperature) + 0.092287 (Medium pH) + 0.000099 (Incubation temperature) (Medium pH) - 0.003517 (Incubation temperature)^2 – 0.010483 (Medium pH)^2 (1)

Besides, the interaction of parameters was classified as A: B (Incubation temperature: Medium pH) and showed in specific via 3D surface plot. Figure 1 illustrated the 3D response surface plot of enzyme activity as a function of incubation temperature and medium pH. From Figure 1, it clearly shown that the optimized condition of the incubation process for maximum xylanase production for both parameters: incubation temperature and medium pH. The 3D response surface plot was observed to project upwards to a region of red colour. This designated that the optimized condition was achieved through this study. The optimized condition was identified allocated at point where the incubation temperature at 32.5°C and medium pH at 5.0 based on the analysis interpreted from the software. From the data interpreted between the interaction of two parameters shown in response surface plot, the maximal xylanase production was recorded at 0.5237 U/mL. Contrarily, the lowest xylanase production concentration is at 0.4000 U/mL.

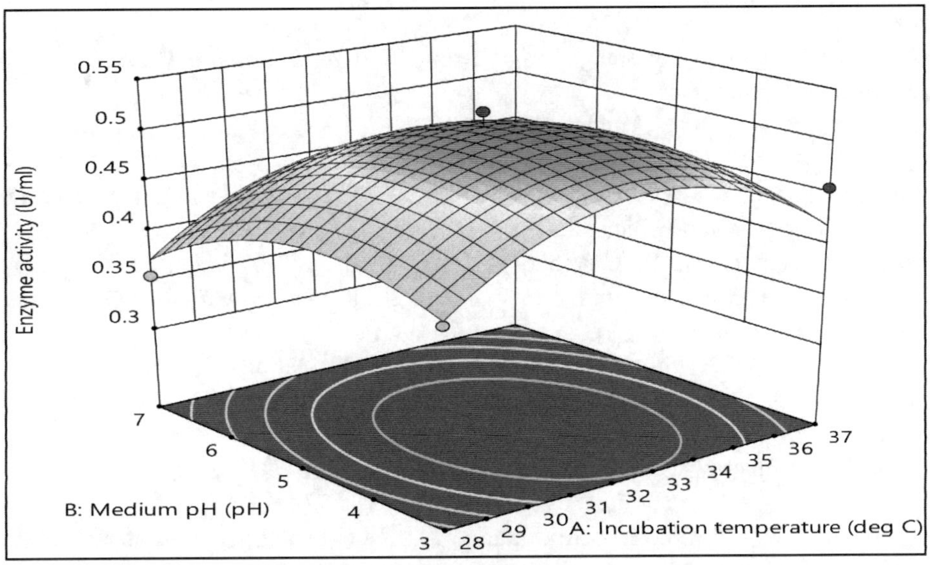

Figure 1: 3D response surface plot of enzyme activity as a function of parameter Design point above predicted value ●; design point below predicted value ○) (0.336776 ▬ 0.523667 colour points by value of enzyme activity

Since the aim of optimization study was to find the conditions which gave the maximum enzyme activity production. Through the analysis of CCD in RSM, the goals for both the incubation temperature and medium pH were selected as "in range" condition, whereas, for enzyme activity, the goal was selected based on maximum in order to achieve the highest xylanase production. From the goals set, the software generated the optimum condition with incubation temperature at 32.67°C and medium pH of 4.56 that predicted will produced 0.5118 U/mL of enzyme activity with desirability of 0.936 from the incubation result.

Hence, validation test was carried using the given optimum condition for incubation to produce xylanase by *A. niger*. A triplicate of optimum condition was done to increase the accuracy of result obtained. The actual enzyme activity was compared with the predicted enzyme activity for the error existed. In the experiment, 0.5638 U/mL was obtained under the stated optimum condition. The percentage error between actual enzyme activity (0.5638 U/mL) and predicted enzyme activity (0.5118 U/mL) is 10.16 % which was considered low. Thus, the result obtained through actual enzyme activity under optimum condition for xylan as production was compatible with the predicted production. This indicated that the optimization achieved in this study was reliable.

4. Conclusion

In this study, the factors manipulated to be optimized through CCD under RSM were inclusive of incubation temperature and medium pH. Based on the analysis, the optimum condition for xylanase production from *A. niger* was acknowledged at temperature of 32.67°C with a medium pH of 4.56 whereby the agitation speed and incubation period were statically set at 150 rpm and 5 days throughout the optimization process, respectively. From the validation done, the maximized xylanase activity was obtained at 0.5638 U/mL. Since through this research an optimized xylanase production condition via *A. niger* was established. Hence, future researches could concentrate on substitution of the model xylan with the lignocellulosic waste to be consecutively used in bioethanol production. This alternative shall best suit the xylanase application in reality, besides reduce the cost spending on xylan.

5. Reference

[1] Shankar U, Uday P, Choudhury P and Kanti T 2016 *International Journal of Biological Macromolecules* **82** 1041–54

[2] Meddeb-mouelhi F, Kelly J and Beauregard M 2014 *Enzyme and Microbial Technology* **66** 16–9

[3] Song Y, Jin E, Song C, Hoon C, Park B and Bae H 2019 s *Renewable Energy* **139** 1281–9

[4] Singh G, Singh H, Kaur S, Bansal S and Kaur S 2011 *Industrial Crops & Products* **34** 1160–7

[5] Modi H A, Acharya P B, Chabhadiya S B, Shah A J, Acharya D K, Acharya, P. B. Chabhadiya S B, Shah A J and Acharya D K 2011 *International Journal of Biotechnology and Biosciences* **1** 423–30

[6] Oliveira P, Cavalieri N, Guimarães D A, Carvalho S De, Betini J H, Marchetti C R, Fonseca F, Lourdes M De, Moraes T De, Marques M R and Giannesi C 2015 *African Journal of Biotechnology Full* **14** 3368–73

[7] Ibrahim D, Haritharan V, and Lim S H 2015 *World Journal of Biological Chemistry* **26** 265–71

[8] Ajijolakewu A K, Leh C P, Wan Abdullah W N, Lee C K, Peng C, Nadiah W and Abdullah W 2017 *Biocatalysis and Agricultural Biotechnology* **11** 239–47

[9] Bailey M J, Biely P and Kaisa P 1992 *Journal of Biotechnology* **23** 257–70

[10] Dinarvand M, Rezaee M and Foroughi M 2017 *Brazilian Journal of Microbiology* 1–15

[11] Shafaghat H, Najafpour G, Sirous-rezaei P, Sharifzadeh M and Branch A 2010 S *Chemical Industry & Chemical Engineering Quarterly* **16** 199–206

Acknowledgement

The author would like to acknowledge the support from the Fundamental Research Grant Scheme (FRGS) under a grant number of FRGS/1/2017/TK02/UNIMAP/02/7 from the Ministry of Education Malaysia. Furthermore, the authors are also grateful to Universiti Tunku Abdul Rahman for the financial aid through the UTAR Research Fund (UTARRF) with a project number IPSR/RMC/UTARRF/2019-C1/T05.

IOP Conf. Series: Materials Science and Engineering **855** (2020) 012006 doi:10.1088/1757-899X/855/1/012006

Kinetic Study of Steam Gasificaiton of Palm Kernel Shell Char

Nuttapol Lerkkasemsan

Department of Chemical Engineering, Faculty of Engineering
King Mongkut's Institute of Technology Ladkrabang, Ladkrabang, 10520 Thailand

Nuttapol.le@kmitl.ac.th

Abstract. In this work, we investigate the reaction mechanism of steam gasification reaction of palm kernel shell char. The experiment results which are used in this work are carried on at temperature of 1123, 1173, and 1223K. The isothermal steam gasification of palm kernel shell char is described by mathematic models including volumetric model (VM), grain model (GM), random pore model (RPM), and modified random pore model (mRPM). From results, we found that the RPM model can predict gasification of palm kernel shell char better than other models. The RPM model considers the overlapping of pore surfaces, which results in the reduction of surface area available for the reaction. The activation energy of gasification reactivity of palm kernel shell char is 152.63 kJ/mol for RPM model. The coefficient of determination of RPM model are 0.9920, 0.9995, and 0.9907 at 1123, 1173, and 1223 K respectively.

1. Introduction

Nowadays, palm oil production is a big industry in Thailand. The increasing of biodiesel demand in Thailand results in higher production of palm oil. Waste from palm oil production also increases as the industry grown. Palm kernel shell waste is also one of the wastes from palm oil production. When the waste from industry is left to environment it would affect to life and ecology. So, we have to search for alternative to solve or alleviate this problem. Activated carbon from palm kernel shell is considered as the alternative for this problem since it can be used in many ways. It consists of carbon that obtained from the activation of char. normally, activated carbon is used for adsorption process because of its high degree of microspore and high surface area.

Activated carbon can produce from two step reaction. Firstly, pyrolysis of biomass to obtained raw materials that carbon rich. Secondly, gasification reaction of char to eliminated contaminant in the pore. Generally, biomass gasification reaction is very complex involves a complex series of chemical reaction. So, it is difficult to find the mathematic model to describe the gasification reaction. The common mathematic model for describe the gasification reaction is gas-solid reactions model that involved four phenomena which are external mass transfer, pore diffusion, adsorption and desorption, and chemical reaction. The mathematics models for this work include volumetric model, grain model, random pore model, and modified random pore model. These models differ in term of primary assumption such as overlapping or non-overlapping of population of grain or pore on the solid. This research will report the suitable model and mechanism for steam gasification of palm kernel shell at different temperature.

2. Theory

2.1. Gasification reaction

Char gasification is two step including pyrolysis and gasification. The pyrolysis involved the evolution of compound of low molecular weight at temperatures between 300 – 500 oC. These compounds are mainly tarring and non-condensable gases. Normally, the pyrolysis residue remains about 70% of original mass. Usually, the gasification reaction is slower than pyrolysis reaction. So, only char gasification reaction is considered for this study. The main reaction for gasification reactions are shown following.

$$C + CO_2 \leftrightarrow 2CO \qquad \Delta H = +172 \quad kJ/mol \tag{1}$$

$$C + H_2O \leftrightarrow CO + H_2 \qquad \Delta H = +131 \quad kJ/mol \tag{2}$$

$$CO + H_2O \leftrightarrow CO_2 + H_2 \qquad \Delta H = -41 \quad kJ/mol \tag{3}$$

$$C + O_2 \rightarrow CO_2 \qquad \Delta H = -394 \quad kJ/mol \tag{4}$$

$$C + 2H_2 \leftrightarrow CH_4 \qquad \Delta H = -75 \quad kJ/mol \tag{5}$$

From the reaction, reaction (1) and (2) are endothermic reaction and they are considered as the important reaction for gasification reaction. The reaction (3) is exothermic reaction which provided energy for reaction (1) and (2). The shift reaction (4) occurs during high steam concentration and reaction (5) is more effective at high pressure. When gasification proceeds, the burn-out rate is used to determine the gasification rate.

$$r = \frac{1}{W}\frac{dW}{dt} = \frac{1}{1-X}\frac{dX}{dt} \tag{6}$$

$$X = \frac{W_0 - W}{W_0} \tag{7}$$

Where r is normalized gasification rate
t is time
W is the char mass at time t
X is char conversion
W_0 is Initial mass of char

$$r' = \frac{-1}{W_0}\frac{dW}{dt} = \frac{dX}{dt} \tag{8}$$

In this case, r' is consider as gasification rate

2.2. Volumetric model [1],[2]

Volumetric model considers a gas-solid reaction take place on a porous particle of general stoichiometry.

$$A(gas) + bB(solid) \rightarrow product \tag{9}$$

Accounting for the effect of solid lost on reaction rate, it may describe by using simple power law rate. For this reaction, the reaction rate is described in first order reaction:

$$-r_A = kC_AC_B \tag{10}$$

where k is rate constant base on unit volume of bulk solid.

Assumption in this mathematic model is consist of 1. The structure change of solid due to reaction is negligible.
2. The system for this reaction is isothermal reaction.
3. Pseudo-steady state approximation is valid.
4. Gas film resistance is negligible.
The conversion rate can be express as followed:

$$\frac{dX_B}{dt} = bkC_A\eta(1-X_B) \tag{11}$$

2.3. Grain model[3],[2]

Grain model is involving heterogeneous gas-solid reaction with moving the boundary. For the grain model, the solid phases consist of unreacted core surrounded by the reacted shell. When the reactions are proceeding, the reacted zone will be expanded and the diffusion of reactant or product through this region may be one of rate limiting step.

The non-react core model for this research is grain model. This model assumes that reacting char particles are considered as spherical grains whose radius decrease as the gasification reaction proceed. The reaction rate for this model expression are.

$$X = 1-(1-kt/3) \tag{12}$$

$$\frac{dX}{dt} = k(1-X)^{2/3} \tag{13}$$

2.4. Random pore model [2],[4]

The random pore model is proposed by Bhatia and Perlmutter. The random pore model can be applied to gasification reaction. The random pore model assumes that there is random overlapping of pore surface which reduced the available for reaction. The equation for this model is:

$$X = 1-\exp\left[-\tau\left(1+\frac{\tau\psi}{4}\right)\right] \tag{14}$$

$$\frac{dX}{dt} = k(1-X)\sqrt{[1-\psi\ln(1-X)]} \tag{15}$$

and

$$\tau = \frac{K_s C^n S_0 t}{1-\varepsilon_0} \tag{16}$$

where S_0 is initial surface area at t = 0
and S_0 can find from pore volume distribution (V_0) of char by equation below.

$$S_0 = \int_0^\infty \frac{V_0(r)}{r}dr \tag{17}$$

where V_0 is pore volume distribution (m^2/g) determine by CO_2 at 273 K.

$$\psi = \frac{4\pi L_0(1-\varepsilon_0)}{S_0^2} \tag{18}$$

where ψ is a structural parameter
L_0 is pore length at t = 0
ε_0 is char porosity at t = 0

2.5. Modified random pore model [2],[5]

The modified random pore model is developed from random pore model, it different from the random pore model by added the conversion term to random pore model. It can be written as:

$$\frac{dX}{dt} = k(1-X)^n\sqrt{[1-\psi\ln(1-X)]} \tag{19}$$

where n is the dimension less power.

3. Result and Discussion

In this work, we employ the experimental result from Chang et al.[6] The reaction was done at three different temperature of 1123, 1173, and 1223 K. In their work, the material was heated from 298K to the designed temperature at a rate of 20°C/min. The steam of 5g/h was used as the gasifying agent.

At first, we employ volumetric model (VM) to simulate gasification reaction of palm kernel shell char with steam. The result is shown below:

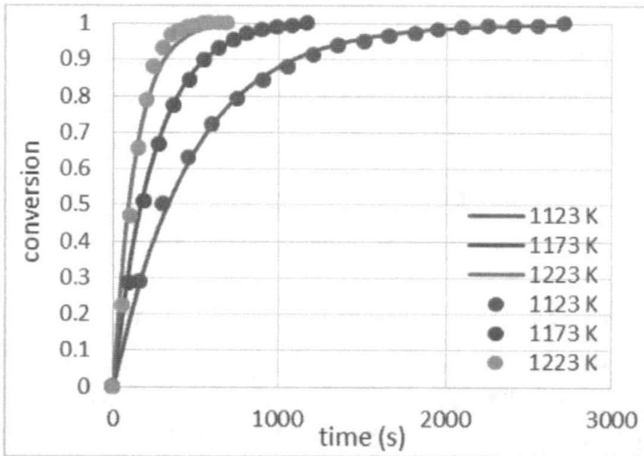

Figure 1. Simulation of steam gasification of palm palm kernel shell char at temperature of 1123 K 1173 K and 1223 K using VM model vs experimental results.

Table 1. VM model parameter results

Parameter	Result		
	1123 K	1173 K	1223 K
k (1/s)	2.12×10^{-3}	4.19×10^{-3}	7.58×10^{-3}
R^2	0.9987	0.9961	0.9831
Average R^2	0.9926		
Ea (J/mol)	145461.74		
A (1/s)	12450.30		

From Figure 1 and Table 1, the VM model seems to work at 1123 and 1173K better than 1223 K. However, the conversion of VM model is lower than experimental data. The results show that VM model is working well for reaction at temperature range of 1123 to 1223 K.

The next model that we are using in this simulation is grain model (GM) model.

Table 2. GM model parameter results

Parameter	Result		
	1123 K	1173 K	1223 K
k (1/s)	1.31×10^{-3}	3.14×10^{-3}	5.44×10^{-3}
R^2	0.9382	0.9953	0.9928
Average R^2	0.9754		
Ea (J/mol)	162896.20		

A (1/s)	51585.71

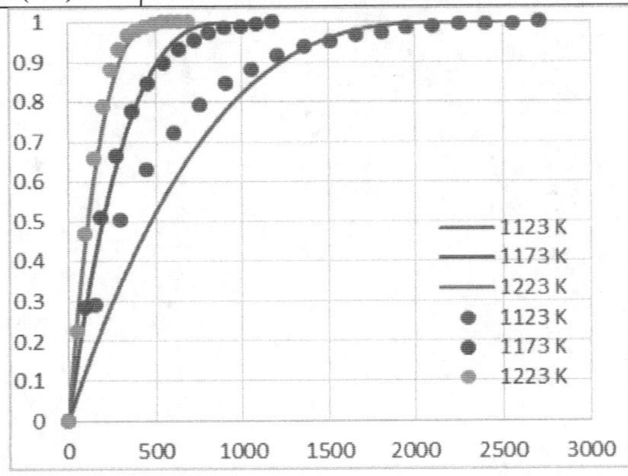

Figure 2. Simulation of steam gasification of palm kernel shell char at temperature of 1123 K 1173 K and 1223 K using GM model vs experimental results.

From Figure 2 and Table 2, the conversion simulated from VM model is better than GM model. GM model can predict conversion at higher temperature (1173 and 1223 K) better than at lower temperature (1123 K)

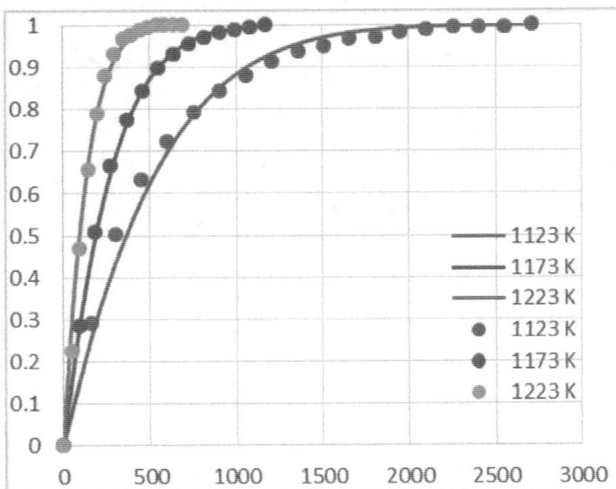

Figure 3. RPM Model for steam gasification of palm kernel shell char at 1123 K 1173 K and 1223 K vs experimental results.

Table 3. RPM model parameter results

Parameter	Result		
	1123 K	1173 K	1223 K
k (1/s)	1.76×10^{-3}	3.40×10^{-3}	6.76×10^{-3}
ψ	0.5025	0.5025	0.5025
R^2	0.9920	0.9995	0.9907
Average R^2	0.9941		
Ea (J/mol)	153293.53		
A (1/s)	23482.25		

From Figure 3 and Table 3, RPM model seems to work well at temperature range from 1123 to 1223 K. The coefficient of determination for RPM model is higher than 0.99 for those three temperatures.

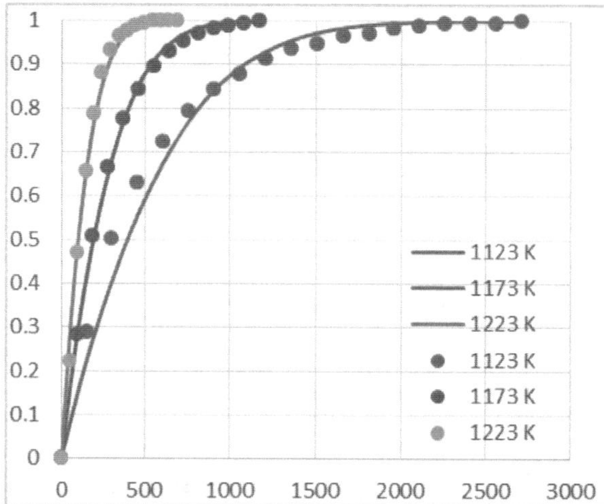

Figure 4. mRPM Model for steam gasification of palm kernel shell char at 1123 K 1173 K and 1223 K vs experimental results.

Table 4. mRPM model parameter results

Parameter	Result		
	1123 K	1173 K	1223 K
k (1/s)	1.54×10^{-3}	3.12×10^{-3}	5.86×10^{-3}
ψ	0.8025	0.8025	0.8025
n	0.9750	0.9750	0.9750
R^2	0.9818	0.9983	0.9956
Average R^2	0.9919		
Ea (J/mol)	152628.41		
A (1/s)	19448.01		

From Figure 4 and Table 4, mRPM model work well at temperature of 1173 and 1223K. However, the model is a little bit off at the temperature of 1123 K.

From the result of four models, the rpm model seems to work well for every temperature. Therefore, the mechanism of steam gasification should follow RPM mechanism model. During the reaction, there is random overlapping of pore surface which reduced available area for reaction.

4. Conclusion

In this research, we investigate the reaction mechanism of steam gasification reaction of palm kernel shell char. The experiment results which are used in this work are carried on at temperature of 1123, 1173, and 1223K. There are four reaction models used in this work. Models are volumetric model, grain model, random pore model, and modified random pore model. From results, the random pore model gives better results at three temperatures than other three models. That means during the reaction, there is random overlapping of pore surface which reduced available area for reaction. The activation energy of gasification reactivity of palm kernel shell char is 152.63 kJ/mol for RPM model. The coefficient of determination of RPM model are 0.9920, 0.9995, and 0.9907 at 1123, 1173, and 1223 K respectively.

5. References

[1] S. Kimura, J.-I. Nakagawa, S. Tone, and T. Otake, "The Volume Reaction Model Based on a Second Order Rate Equation and Its Application to a Gas-Solid Reaction," *J. Chem. Eng. Japan*, vol. 14, no. 3, pp. 190–195, 1981.

[2] G. Wang *et al.*, "Experimental and modeling studies on CO2 gasification of biomass chars," *Energy*, vol. 114, pp. 143–154, 2016.

[3] H. Y. Sohn and J. Szekely, "A structural model for gas-solid reactions with a moving boundary-IV. Langmuir-Hinshelwood kinetics," *Chem. Eng. Sci.*, vol. 28, no. 5, pp. 1169–1177, 1973.

[4] S. K. Bhatia and D. D. Perlmutter, "A Rondom Pore Model for Fluid-Solid Reactions : 1 . Isothermal , Kinetic Control," *AIChE J.*, vol. 26, no. 3, pp. 379–386, 1980.

[5] J. Zhang, G. Wang, J. Shao, and H. Zuo, "com A Modified Random Pore Model for the Kinetics of Char Gasification," *Bioresources*, vol. 9, no. 2, pp. 3497–3507, 2014.

[6] G. Chang *et al.*, "The lignin pyrolysis composition and pyrolysis products of palm kernel shell , wheat straw , and pine sawdust," *Energy Convers. Manag.*, vol. 124, pp. 587–597, 2016.

Acknowledgment

I would like to express my gratitude to the National Science and Technology Development Agency for funding.

IOP Conf. Series: Materials Science and Engineering **855** (2020) 012007 doi:10.1088/1757-899X/855/1/012007

Ultrathin Mica and Graphite Cantilevers Enhanced Photoacoustic Spectroscopy – towards Modelling of Acousto-mechanical Properties

Suchánek Jan[1], Dostál Michal[1], Janda Pavel[1], Zelinger Zdeněk[1], Chalupský Jaromír[2], Wild Jan[3]

[1]J. Heyrovsky Institute of Physical chemistry, Academy of Sciences of the Czech Republic, Prague, Czech Republic
[2]Institute of Physics, Academy of Sciences of the Czech Republic, Prague, Czech Republic
[3]Faculty of Mathematics and Physics, Charles University, Prague, Czech Republic

zdenek.zelinger@jh-inst.cas.cz

Abstract. In this article, we present the concept of probing of the acousto-mechanical properties (e.g. spring constant, resonator quality factor) of thin 2D or 2D like samples (i.e., samples with one dimension at least 2 orders of magnitude smaller than the others). Samples prepared from highly ordered pyrolytic graphite and muscovite with thickness under 1 μm and other dimensions in the mm range are used for this purpose. These samples are employed in photoacoustic spectroscopy as the pressure sensitive elements and its movement is monitored by a He-Ne laser beam reflected from the surface of these samples onto a position sensitive detector. The concept includes the following steps: 1) development of the method of preparation of thin samples from layered materials 2) development of the experimental approach for data collection 3) derivation of theoretical behavior of photoacoustic response of the samples and 4) development and application of a model on experimentally acquired data.

1. Introduction

Micro and nanomechanical 2D materials show extreme sensitivity to mass and surface changes and can thus be exploited as sensitive chemical and biological sensors with the ultimate sensitivity down to individual molecules. Such materials can be used as nanomechanical resonators with resonance frequencies above 1 GHz and with high quality factors (10^3 - 10^5) [1,2]. Its supreme sensitivity to mass was exploited for the measurement of physical quantities [3–5] Beside ultrasensitive mass detection , resonators with high resonance frequencies can be applied in signal processing and communication technologies.

Layered materials such as highly ordered pyrolytic graphite (HOPG) and multilayer muscovite (mica, MLM) are ideal starting materials for the preparation of ultrathin (down to single layer) 2D materials. Its thickness can be simply reduced by exfoliation of the basal planes. Bunch et. al. demonstrated methods for the characterization of mechanical properties (resonance frequency, spring constant, quality factor) for nanoresonators prepared from graphene sheets [6]. The resonators are actuated either optically or electrically and are detected optically by interferometry. High quality factors were demonstrated for these nanoresonators with resonant frequencies in the MHz range. Recently, frequency tracking through resonant frequency comb was reported [7].

The extreme sensitivity of nano/micromechanical elements to external force can be exploited in photoacoustic spectroscopy (PAS). PAS is very sensitive detection method, where the periodically interrupted radiation is converted to heat when absorbed by a substance (gas, solid or liquid) and subsequently converted to acoustic waves, which can be detected by a microphone. An optical microphone was proposed by de Paula et.al. – a pressure sensitive element which motion can be sensed optically with a position sensitive detector [8]. Wilcken and Kauppinen demonstrated a 5 micrometer thin silicon cantilever used as an optical microphone and showed both theoretically and experimentally that such microphone exceeds conventionally used (electret and condenser) highest-class microphones [9]. Cantilever enhanced photoacoustic spectroscopy (CEPAS) is very promising sensitive trace gas detection method [10–12].

In our previous work, we suggested that thin graphite and multilayer graphene sheets can be utilized in PAS as pressure sensitive elements – optical microphones [13–15]. In our first work on this topic [13], we tested multilayer graphene sheets prepared from multilayer graphene through the micromechanical cleavage of basal plane highly ordered pyrolytic graphite. The sheets with thickness ~ 100 nm were glued to a glass substrate with a circular opening (r ~ 1 mm). We showed the feasibility of using these elements for PAS, although the sensitivity was inferior to a commercial highest-class microphone. Through the development of the preparation method, we prepared a multilayer graphene with cantilever-like geometry that surpassed the conventional microphone [15]. Nevertheless, we have found that the non-uniform structure of the cantilever caused by the presence of grain boundaries of the graphene sheets limits its elastic properties and thus its performance as an acoustic resonator. Next, we focused on multilayer mica as starting material for cantilever preparation [16]. It seems that the structure of mica sheets prepared by exfoliation of basal planes is more uniform when compared to the graphene sheets prepared by the same method from HOPG. Thanks to the progress in cantilever preparation, the sensitivity was further increased, although the cantilevers suffered from long term instabilities when used as a part of an optical microphone.

We have also suggested, that PAS can serve as a method for probing the mechanical properties of thin samples like graphene and mica sheets and cantilevers [14]. In this article, we explore this idea into more depth.

2. Experimental

2.1. Photoacoustic spectroscopy setup

A modified setup for photoacoustic spectroscopy was developed to test ultrathin samples (Fig. 1). A quantum cascade (QC) laser emitting between 7.996 – 8.009 μm / 1249 - 1251 cm^{-1} (QD4580, Thorlabs, USA) was used as an excitation source for PA signal generation. The QC laser was installed in a laser mount with temperature stabilization (Arroy 242, Arroyo Instruments, USA) and controlled by a laser current/temperature controller (ITC4005QCL, Thorlabs, USA). The laser beam was collimated by an aspheric lens, periodically interrupted by a chopper and focused inside a custom-made photoacoustic cell with inner diameter of 6 mm and a length of 245 mm (length can be adjusted via changeable arms). Part of the laser beam was reflected by a beam splitter onto nitrogen cooled InSb detector with preamplifier (Graseby Infrared) to monitor the laser beam. The photoacoustic cell is an important part of the setup and was designed to enable easy exchange of tested materials (graphene/mica membranes and cantilevers). The samples were placed on a glass substrate. The movements of the cantilever ends induced by acoustic waves were actuated by a He–Ne laser beam that was reflected from the cantilever ends onto a quadrant detector (a red-enhanced quad-cell silicon photodiode, SD 085-23-21-021, Laser Components). The PA cell has a place for mounting both the samples and conventional microphone and is made symmetrical, so the comparison of the samples and the microphone (Brüel and Kjaer, condenser microphone, type 4144) is possible. There is a small opening in the middle of the cell leading to the microphones. The detail of the custom-made PA cell can be found in the paper [16]. The signals (waveforms) from the quadrant cell (mica and graphite cantilever), microphone, InSb detector (QC laser - excitation source of photoacoustic signal) were collected via 4-channel digital oscilloscope (DSO5054A, Agilent Technologies, USA).

2.2. Cantilever preparation

The preparation of multilayer graphene/muscovite cantilever is described in detail in our previous paper [16]. Briefly, the sample was prepared by the Scotch tape method (by peeling the basel planes) and then cutting them to desired shape. The surface was covered by thing AU film (50 – 100 nm) for better reflexiveness. The cantilever was circular with 13 mm diameter and thickness under 1 μm.

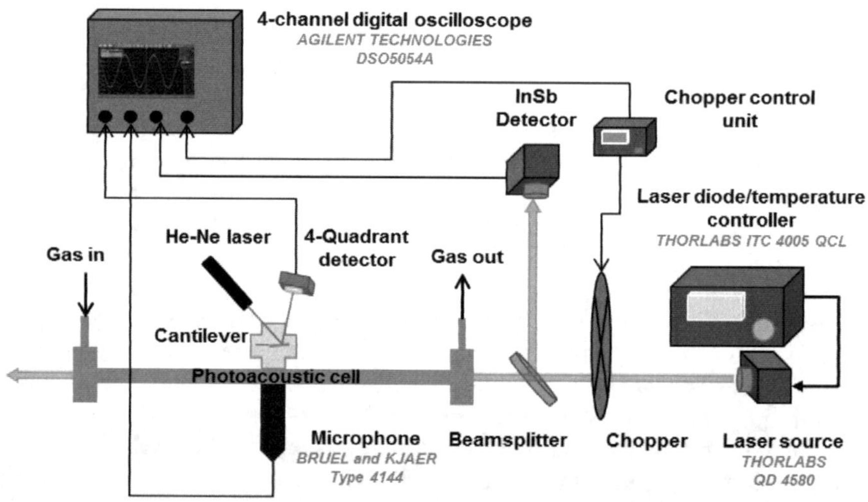

Figure 1. The photoacoustic experimental setup.

3. Results and discussion

3.1. Experimental approach for data collection

To get the relevant inputs for a photoacoustic model, the experimental approach described above was developed. This setup enables the collection of the signals from a) the 4-quadrant detector (i.e. – the cantilever/membrane movement) b) the microphone c) an optical detector (profile of the excitation laser beam) and d) reference signal from the chopper control unit (information about phase shift). The example of collected signals is on Fig. 2.

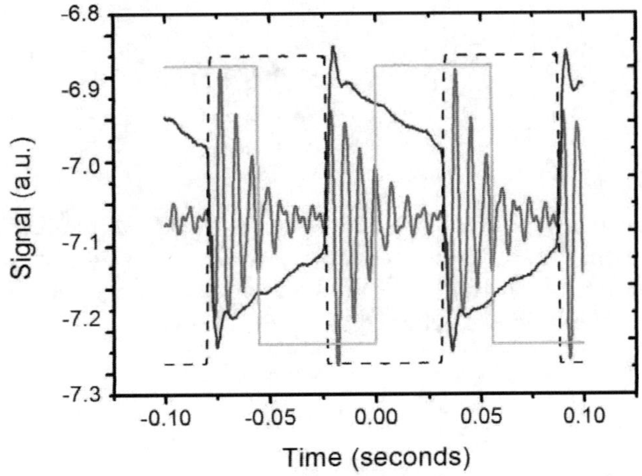

Figure 2. Typical experimental results. 9 Hz modulation. Cantilever movement (red), condenser microphone movement (blue), excitation laser profile (black dash) and chopper reference signal (green).

To test the developed (acousto-mechanical) models thoroughly, the response of the samples (cantilevers, membranes) to photoacoustic pressure can be tested at various conditions. The change of carrier gas influences the sound velocity and thus the PA cell resonant frequencies, which can be monitored by changing the modulation frequency of the source of radiation. The acoustic properties of the PA cell can be varied (different PA cell length, closed/open gas inlet and outlet arrangement).

3.2. Derivation of theoretical photoacoustic behavior

This section describes the theoretical aspects of PA signal behavior, as derived by the authors. The derivation are not shown here in full details, as it is beyond the scope of this article, but only the main starting points and results are highlighted. For more details, please see [17][10]

When investigating the process of acoustic wave formation in a PA cell, it is necessary to observe all influences on the volume element of the gas inside the cell. Pressure changes caused by the absorption of laser radiation create an acoustic field inside the PA cell.

Photoacoustic signal generation can be divided into these steps: 1) periodic vibration/electronic excitation of molecules into higher state by radiation source - 2) vibrational-transational relaxation followed by heating of the gas and finally 3) the formation of acoustic wave. The acoustic and thermal wave generation inside of the PA cell can be modelled by applying the laws of thermodynamics a fluid mechanics. The time-depended pressure changes inside the PA cell can be theoretically expressed as follows:

$$\frac{dP}{dt} = \frac{2}{3} h\nu \frac{n_1}{\tau_c} - (P - P_0) / \tau_1 \qquad (1)$$

where P_0 is the equilibrium pressure inside the cell without radiation, h is the Planck-constant, v is the frequency where the sample absorbs, n_1 (m^{-3}) is the number of molecules in unit volume in higher energetic state with absorption frequency equal to the frequency of the radiation, τ_c is the collision-controlled lifetime of the higher energy state and τ_1 is the time of photoacoustic delay which depends on the thermal properties of the gas.

If we consider the time dependence of n_1 in eq. 1 (not shown here), we can write the pressure inside closed (non-flow) PA cell as a function of time:

IOP Conf. Series: Materials Science and Engineering **855** (2020) 012007 doi:10.1088/1757-899X/855/1/012007

$$P(t)=\begin{cases} P_0+A\left(1-e^{-\frac{t}{\tau_1}}B\right) & t\in\left[0,\frac{T}{2}\right] \\ P_0+ABe^{-\left(t-\frac{T}{2}\right)/\tau_1} & t\in\left[\frac{T}{2},T\right] \end{cases} \tag{2}$$

where A and B are parameters. The pressure changes corresponding to the first harmonic in the Fourier series, which represents the time dependence of the pressure change, are most pronounced when interacting with the pressure sensor. If we take the Fourier series of eq. 2 and take into account its first harmonic frequency and simplify it, we get to eq. (3)

$$S_1(\omega)=\frac{\sqrt{2}}{\pi^2}(\gamma-1)\frac{NW\sigma}{R^2\left(1+\omega^2\tau_1^2\right)^{1/2}}Q_c(\omega)\tau_1 \tag{3}$$

This equation describes the dependence of PA signal on the concentration of absorbing molecules N, on the radiation source power W, on the absorption cross-section σ of the absorbing molecules, on the time-delay of PA response τ_1, on the modulation frequency $\omega=2\pi f$, on the PA cell radius R and on the PA cell quality factor Q_c.

The movement of the pressure actuator (i.e. membrane, cantilever) can be described by the second order one-dimensional eq. (4):

$$\sigma_m A_m\frac{d^2x_m}{dt^2}+D\frac{dx_m}{dt}+Gx_m=A_m p_{\sim} \tag{4}$$

where σ_m is the area density of the actuator, x_m is the displacement of the actuator from equilibrium, $p_{\sim}$ is the actuating pressure, A_m is the area of the actuator, $D\ dx_m/dt$ is the damping force and Gx_m is the restoring force acting on the actuator.

The behavior of the gas inside of the cell can be approximated by the equation of forced damped harmonic oscillator. Its main characteristics are the spring constant k, the damping constant B_c and mass (of the gas inside of the cell) – equation 5.

$$\rho_0 V\frac{d^2x_c(t)}{dt^2}=-kx_c(t)-B_c\frac{dx_c(t)}{d_t}+\frac{2}{3}\nabla(\psi-\psi_0) \tag{5}$$

where V is the volume of the PA cell and $x_c(t)$ is the displacement of the volume element inside of the cell from equilibrium, ψ is the average transitional energy of the molecules in unit volume and ψ_0 is the equilibrium value of ψ.

In case of the connection of the PA cell with flow system (which is the common case for photoacoustic setup), we have derived following equation:

$$p_{\sim}=p_{id}\left[1-2c_1te^{-c_1^2t}I_1(t)\right] \tag{6}$$

where $p_{\sim}$ is the oscillating part of the pressure inside of the PA cell, c_1 is a constant that depends on the density of the gas and dimension of the PA cell, and I_1 is given by eq. 7:

$$I_1(t)=\sum_{i=0}^{\infty}c_1^i\frac{t^{2i+1}}{(2i+1)(i!)} \tag{7}$$

By substitution of the eq. 6 into equations 4 and 5, it is possible to get the second order inhomogeneous equation that describes the time dependence of the PA signal:

$$\ddot{x}_m+2k_m\dot{x}_m+\omega_m^2x_m=F_mp_{\sim}(t) \tag{8}$$

$$\ddot{x}_c+2k_c\dot{x}_c+\omega_c^2x_c=F_cp_{\sim}(t) \tag{9}$$

where

IOP Conf. Series: Materials Science and Engineering **855** (2020) 012007 doi:10.1088/1757-899X/855/1/012007

$$k_m = \frac{\omega_m}{2Q_m}; k_c = \frac{\omega_c}{2Q_c}; F_m = \frac{1}{\sigma_m}; F_c = \frac{A_c}{\rho_0 V}; \omega_m = 2\pi\nu_m$$

$$k_m = \frac{D}{2\sigma_m A_m}; \omega_m^2 = \frac{G}{\rho_0 V}; k_c = \frac{B_c}{2\rho_0 V}; \omega_c^2 = \frac{\kappa}{\rho_0 V};$$

$$\omega_c = 2\pi\nu_c$$

and ω_m and ω_c is the resonance frequency of the actuator and the cell, respectively, Q_m and Q_c is the quality factor of the actuator and cell, σ_m is the area density of the actuator and A_c is the inner area of the PA cell. Through the superposition of the displacement of the volume elements in PA cell x_c, caused by the oscillating pressure $p_{\sim}(t)$ and the response of the pressure actuator x_m, we can derive the time dependent displacement of the photoacoustic response:

$$x(t) = x_m + x_c \tag{10}$$

3.3. *Example of application of developed model*

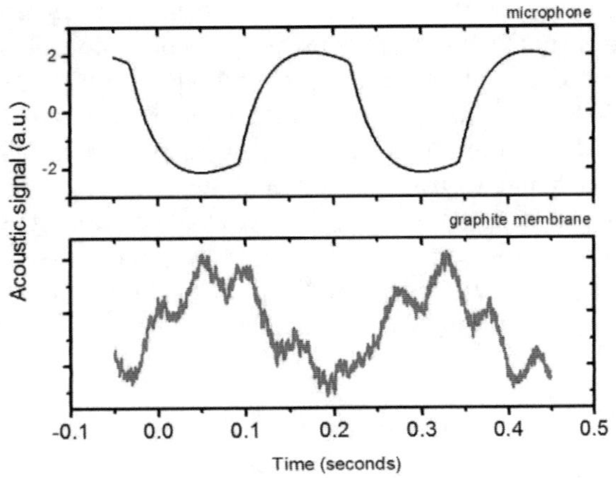

Figure 3. Experimental data of photoacoustic signal.

The equations derived above were used to model the acoustic behavior of the system – i.e. PA cell and cantilever/membrane. The model consists of two coupled forced oscillators, which represent both the generation of acoustic waves in the space of the PA cell and the interaction of pressure changes with the sensor.

To evaluate the model, the experimentally acquired photoacoustic signals (Fig. 3) are compared to the simulated data using the model (Fig. 4).

At a given stage of the model development, the model is able to simulate a PA signal for different PA cell configurations and pressure sensors. This helps to interpret the process of generating of PA signals. The next planned stage of the model development is to fit the theoretical data with the experimental data to obtain data on acousto-mechanical parameters of the materials used.

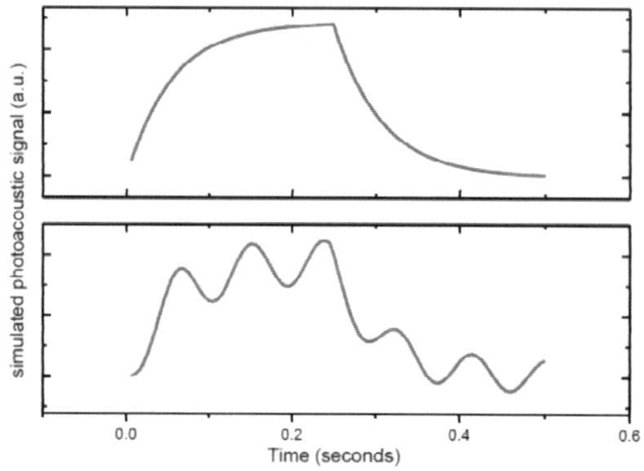

Figure 4. Simulated data of experimental signal.

4. Conclusion
We have developed and tested the approach (both experimental and theoretical) to get the material information related to micro-mechanical levers applicable in photoacoustic spectroscopy.
It includes the development of a method for the preparation of thin samples from carbon and mica layered materials, the development of an experimental approach to data collection using laser photoacoustic detection and the use of the optical microphone and cantilever principle, the derivation of the theoretical behavior of photoacoustic response of samples based on formulation of processes during radiation absorption, generation of thermal changes and creation of acoustic field inside PA cell, further development and application of the model to experimentally obtained data, where the core of the model lies in the coupling of two forced oscillators describing the dynamics of pressure changes in the PA cell space and the interaction with the pressure sensor.

5. References
[1] K. Jensen, K. Kim, A. Zettl, An atomic-resolution nanomechanical mass sensor, Nat. Nanotechnol. 3 (2008) 533–537. doi:10.1038/nnano.2008.200.
[2] K.L. Ekinci, Y.T. Yang, M.L. Roukes, Ultimate limits to inertial mass sensing based upon nanoelectromechanical systems, J. Appl. Phys. 95 (2004) 2682–2689. doi:10.1063/1.1642738.
[3] F. Ricci, M.T. Cuairan, G.P. Conangla, A.W. Schell, R. Quidant, Accurate Mass Measurement of a Levitated Nanomechanical Resonator for Precision Force-Sensing, Nano Lett. 19 (2019) 6711–6715. doi:10.1021/acs.nanolett.9b00082.
[4] G. Gruber, C. Urgell, A. Tavernarakis, A. Stavrinadis, S. Tepsic, C. Magén, et al., Mass Sensing for the Advanced Fabrication of Nanomechanical Resonators, Nano Lett. (2019). doi:10.1021/acs.nanolett.9b02351.
[5] A. Blaikie, D. Miller, B.J. Alemán, A fast, sensitive, room-temperature graphene nanomechanical bolometer, (2018) 1–8. doi:10.1038/s41467-019-12562-2.
[6] J.S. Bunch, A.M. van der Zande, S.S. Verbridge, I.W. Frank, D.M. Tanenbaum, J.M. Parpia, et al., Electromechanical resonators from graphene sheets., Science. 315 (2007) 490–3. doi:10.1126/science.1136836.
[7] A. Ganesan, A. Seshia, Resonance tracking in a micromechanical device using phononic frequency combs, Sci. Rep. 9 (2019) 9452. doi:10.1038/s41598-019-46003-3.
[8] M.H. de Paula, A.A. de Carvalho, C.A. Vinha, N. Cella, H. Vargas, Optical microphone for photoacoustic spectroscopy, J. Appl. Phys. 64 (1988) 3722–3724. doi:10.1063/1.341416.
[9] K. Wilcken, J. Kauppinen, Optimization of a Microphone for Photoacoustic Spectroscopy, Appl. Spectrosc. 57 (2003) 1087–1092. doi:10.1366/00037020360695946.
[10] M. Dostál, J. Suchánek, V. Válek, Z. Blatoňová, V. Nevrly, P. Bitala, et al., Cantilever-Enhanced Photoacoustic Detection and Infrared Spectroscopy of Trace Species Produced by Biomass Burning, Energy and Fuels. (2018). doi:10.1021/acs.energyfuels.8b01021.

[11] T. Tomberg, M. Vainio, T. Hieta, L. Halonen, Sub-parts-per-trillion level sensitivity in trace gas detection by cantilever-enhanced photo-acoustic spectroscopy, Sci. Rep. 8 (2018) 1848. doi:10.1038/s41598-018-20087-9.

[12] T. Tomberg, T. Hieta, M. Vainio, L. Halonen, Cavity-enhanced cantilever-enhanced photo-acoustic spectroscopy, Analyst. 144 (2019) 2291–2296.

[13] Z. Zelinger, P. Janda, J. Suchánek, M. Dostál, P. Kubát, V. Nevrlý, et al., Silicon micro-levers and a multilayer graphene membrane studied via laser photoacoustic detection, J. Sensors Sens. Syst. 4 (2015) 103–109. doi:10.5194/jsss-4-103-2015.

[14] J. Suchanek, M. Dostal, T. Vlasakova, P. Janda, M. Klusackova, P. Kubat, et al., Thin graphite membranes for laser photoacoustic spectroscopy, in: 2015 IEEE Nanotechnol. Mater. Devices Conf. NMDC 2015, 2016. doi:10.1109/NMDC.2015.7439258.

[15] J. Suchánek, M. Dostál, T. Vlasáková, P. Janda, M. Klusáčková, P. Kubát, et al., First application of multilayer graphene cantilever for laser photoacoustic detection, Meas. J. Int. Meas. Confed. 101 (2017) 9–14. doi:10.1016/j.measurement.2017.01.011.

[16] J. Suchánek, P. Janda, M. Dostál, A. Knížek, P. Kubát, P. Roupcová, et al., Photoacoustic spectroscopy with mica and graphene micro-mechanical levers for multicomponent analysis of acetic acid, acetone and methanol mixture, Microchem. J. 144 (2019) 203–208. doi:10.1016/j.microc.2018.08.034.

[17] A. Miklós, Acoustic aspects of photoacoustic signal generation and detection in gases, Int. J. Thermophys. 36 (2015) 2285–2317.

Acknowledgment
The authors are grateful for the financial support from the Czech Science Foundation, Project No. 17-05167S, from the project No. VI20192022127 funded by the Ministry of the Interior of the Czech Republic and from the project ERDF/ ESF "Centre of Advanced Applied Sciences" (No. CZ.02.1.01/0.0/0.0/16019/ 0000778).

IOP Conf. Series: Materials Science and Engineering **855** (2020) 012008 doi:10.1088/1757-899X/855/1/012008

Effect of Substrate Thickness on Horizontal Magnetoresistance

T. Phetchakul[1], Y. Chemthung[1] and A. Poyai[2]

[1] Faculty of Engineering, King Mongkut's Institute of Technology Ladkrabang, Thailand.
[2] Thai Microelectronics Center, Design & Engineering Consulting Service Center (DECC), National Science and Technology Development Agency (NSTDA), Thailand

toempong.ph@kmitl.ac.th

Abstract. This paper presents the effect of substrate thickness on horizontal magnetoresistance that detects the horizontal magnetic field perpendicular to bulk cross section area. The substrate thickness (*T*) concerns directly to the current density distribution and the effective depth (*t*) of current path corresponding to cross section area of current. The horizontal magnetoresistance depends on current distribution which concerns with substrate concentration, length, width and thickness of resistor. It changes by the effective depth of current path change by the unbalance between Lorentz's force and Hall electrical force in vertical direction. From this study, the sensitivity of device increases with the thickness of substrate up to critical substrate thickness and still constant for thickness greater than this critical value. The current distribution is limited when substrate thickness less than a critical value and fully distributes when the thickness is greater than this value. At the same constant current, the longer length of resistor shows the lower current density which needs the longer effective depth of current which causes the critical substrate thickness to be increased accordingly.

1. Introduction

The geometry magnetoresistance is the Hall effect current mode for non-magnetic material such as semiconductor [1-3]. It uses Lorentz's force induced from vertical magnetic field deflect currents to make the current path longer, resulting in higher resistances. It has to design a specific geometry to make induced Lorentz's force greater than induced Hall electrical force for current deflection [4]. The magnetoresistace also occurs in horizontal magnetic field [5]. It occurs as a latent phenomenon in the operation of the device, for example, 5-contact vertical Hall device for detecting horizontal magnetic fields [6]. In addition to the Hall effect caused by horizontal magnetic fields, there is also a horizontal magnetoresistance phenomenon [5, 6]. The mechanism of this effect is not as clear as vertical magnetoresistance because of the asymmetry current path in vertical direction [7]. The mechanism is the unbalance between Lorentz's force and Hall electrical force in vertical with asymmetry current paths of planar resistor [8]. The horizontal magnetoresistance is a geometry magnetoresistance that depends on geometry of device in planar. The thickness of bulk is an important variable and directly related to this phenomenon other than the geometric of device [9, 10]. In this study, the effect of thickness on horizontal is studied by TCAD method [11, 12]. The simulation helps to see the distribution of current density from the magnetic field, which leads to more understanding.

2. Mechanism of horizontal magnetoresistance Introduction

The mechanism of horizontal magnetoresistance can be explained as in figure 1. The horizontal magnetoresistance will occur when the substrate concentration is low [13]. The current is biased by constant current from cc1 to cc2 and electron in n-type substrate flow from cc2 to cc1 in bulk with the effective depth t as shown in figure 1 (b). When the magnetic field is applied in –y direction as shown in figure 1 (a), the induced Lorentz's force F_L acts upon the electron in –z direction. The Hall electric field is induced in the opposite z direction. In this case, the force from Hall electrical field F_H is less than Lorentz's force ($F_L > F_H$) causes the electron current paths are longer deep into bulk in –z direction and increases resistance. On the other hand, the Lorentz's force is induced in z direction when the magnetic field is applied in y direction. In this case, the Hall electrical force is induced into bulk in -z direction and less than Lorentz's force ($F_H < F_L$) and electron current paths are shorter up to surface and the resistance is reduced. The reason of the unbalance of two forces because the current paths in cross section area is asymmetry as in the case of planar. It can be explained as in the equation below

$$R = \rho \frac{L}{A} = \left(\frac{\rho}{t}\right)\left(\frac{L}{W}\right) \tag{1}$$

where R is resistance, ρ is resistivity, L is length of resister and A is cross section area of current. The cross section A can be rewritten into term of width (W) and thickness of current (t) which is appropriate average value that represent of depth corresponding to the cross section area A and resistance change ΔR when applied magnetic field is

$$\Delta R = \rho \left(\frac{L}{W}\right)\cdot\frac{1}{\Delta t} \tag{2}$$

where Δt is the average current depth change from magnetic field. The average current depth t depends on the current distribution profile which is concern with the thickness of substrate T.

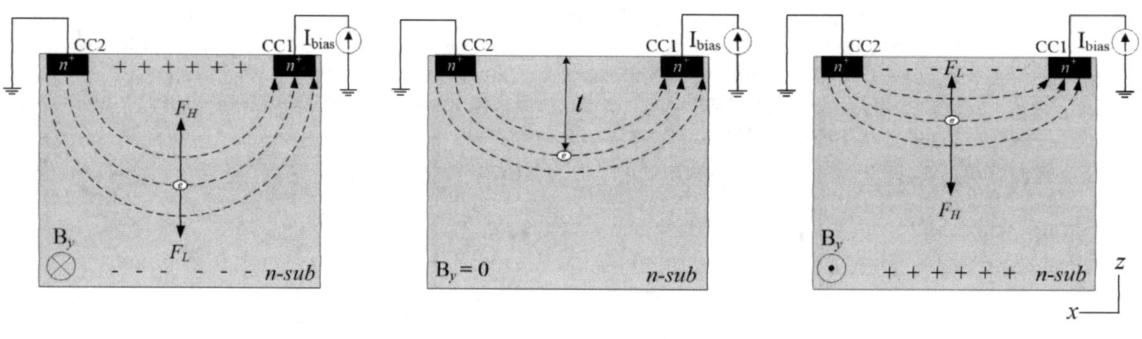

(a) (b) (c)

Figure 1. Cross section of horizontal magnetoresistance at low concentration substrate.

3. Device structure and TCAD

The device and simulation structure is shown in figure 2. The dimensions are width W_d, length L_d, thickness T, contact width W and length between contact L. The substrate is N-type with the low concentration 10^{14} cm^{-3}. The thickness is varied at 5, 10, 15, 20, 30 and 70 μm. The contact width is 5 μm and length L is varied at 5, 10 and 25 μm. The biased current is 1 mA. The tool for this study is by simulation by Synopsis TCAD Sentaurus program. The important semiconductor equations and magnetic model are completely in this program [14].

IOP Conf. Series: Materials Science and Engineering **855** (2020) 012008 doi:10.1088/1757-899X/855/1/012008

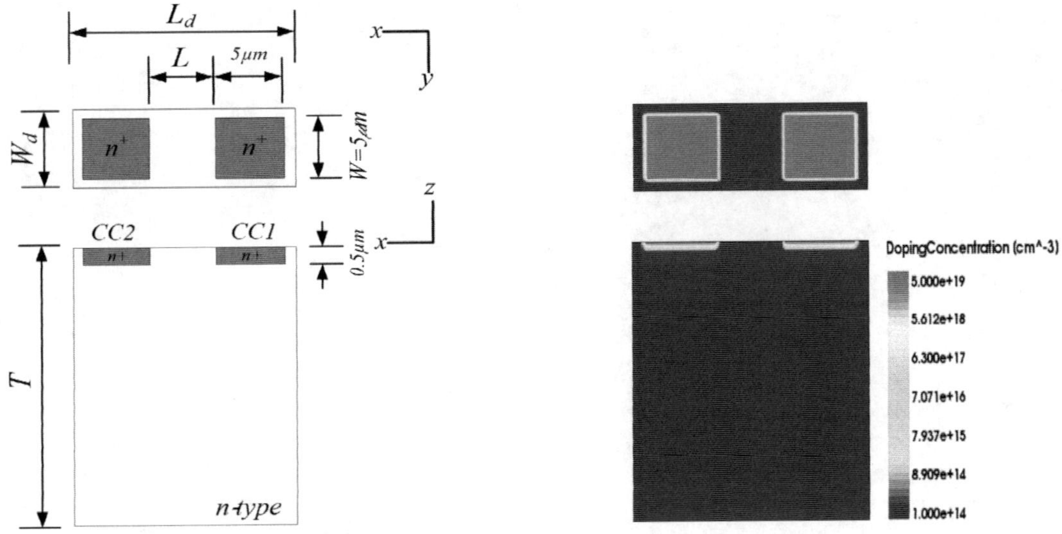

(a) (b)

Figure 2. Resistor structure (a) Device dimensions (b) Simulation structure.

4. Experiments and Discussion

The parameters that we study are substrate thickness T of the length between contacts L. The magnetic field B is applied perpendicular to cross section area in parallel direction. The horizontal magnetic responses that we study are the resistance change (ΔR) and percentage of resistance change (% MR). The definition of them are written as

$$\%MR = \%\frac{\Delta R}{R} = \frac{R(B) - R(0)}{R(0)} \times 100 \tag{3}$$

where $R(B)$ is applied magnetic field resistance and $R(0)$ is non-applied magnetic field resistance. The ΔR and % MR of magnetoresistance are shown in figure 3. The resistance L= 5, 10 and 25 µm, biased current 1 mA and magnetic field 0 – 0.5 T at substrate thickness 5, 10, 15, 20, 30 and 70 µm are in figure 3(a), (b) and (c), respectively. The sensitivity (S) or slope of graph, the resistance change ΔR and % MR increase with substrate thickness up to a critical substrate thickness value and still constant when substrate thickness is greater than this value. The critical thickness of resistor L = 5, 10 and 25 µm are approximately 5, 15 and 30 µm, respectively. All of them show linearly dependence. The non-magnetic field resistance $R(0)$ reduces with the substrate thickness down to the saturation value. Table 1 summarizes $R(0)$, ΔR, %MR and S of all devices. The % MR of longer resister at same constant current is greater than the shorter resistance. The % MR is less than 1 % but if it is considered the resistance value, it is changed up to several kilo ohms for L = 25 µm at 0.5 T by the substrate thickness more than 30 µm.

Figure 4 and 5 show the current distributions of magnetoresistance of the shortest resistor L = 5, and longest resistor in this study L = 25 µm at substrate thickness 5, 10 and 30 µm, respectively. The short resistor in figure 4, the current density is high between the contacts and the long resistor in figure 5, the current density is lower than the short one. The current density depth of short resistor in figure 4 is shallow than the long one in figure 5. It can be said that the current density is reduced according to the length L of resistor for same constant current. At substrate thickness 5 µm, it can be observed that the depth of current is limited by the substrate thickness both of short and longer resistor as shown in figure 4 (a) and (5). At substrate thickness 10 µm, the depth of current of shorter resistor in figure 4 (b) is mostly covered by the thickness of substrate than in the case of longer resistance L = 25 µm. It can be observed that in figure 5 (b) the depth of current is almost covered by substrate thickness but is not complete. At the substrate depth 30 µm in figure 4 (c) and 5 (c), the shorter resistor current depth is

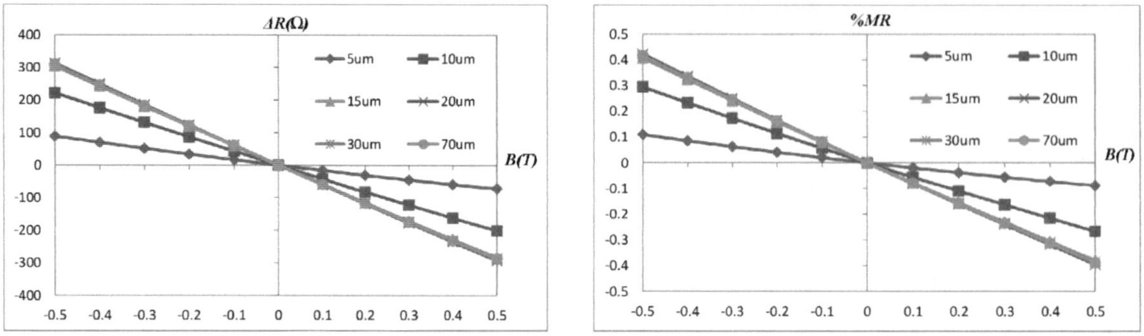

(a) $L = 5$ μm

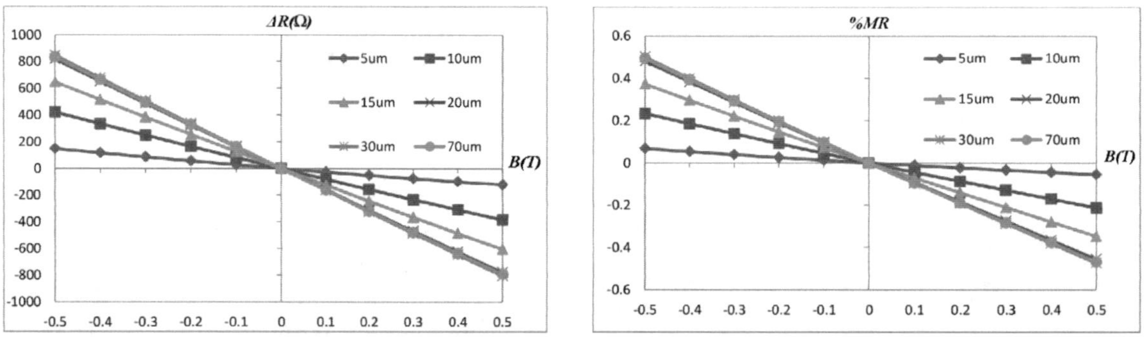

(b) $L = 10$ μm

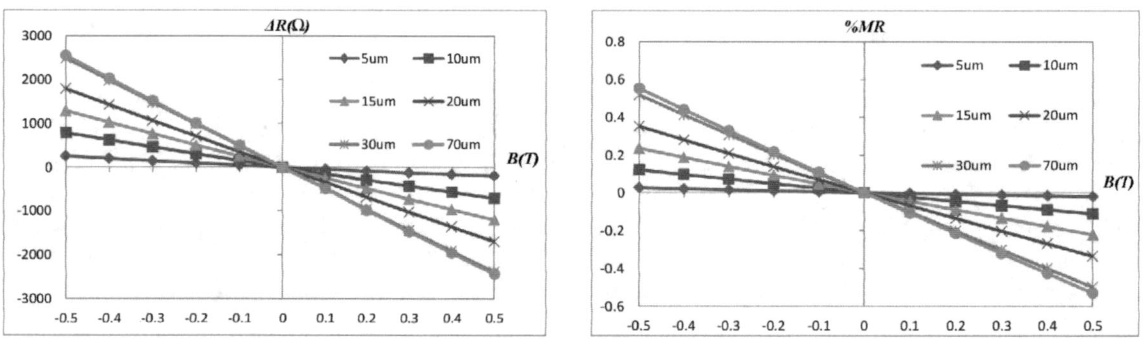

(c) $L = 25$ μm

Figure 3. Magnetoresistance and % *MR* responses of resistor.

Table 1. *R*, Δ*R*, *%MR*, *S* of magnetoresistance

T (*μm*)	*L = 5 μm*				*L = 10 μm*				*L = 25 μm*			
	R (kΩ)	Δ**R** (Ω)	**%MR** (%)	**S** (kΩ/AT)	**R** (kΩ)	Δ**R** (Ω)	**%MR** (%)	**S** (kΩ/AT)	**R** (kΩ)	Δ**R** (Ω)	**%MR** (%)	**S** (kΩ/AT)
5	81.3	88	0.10	17.723	220	147	0.06	29.551	994	254	0.02	50.908
10	75.5	221	0.29	44.368	180	419	0.23	83.900	637	782	0.12	156.58
15	74.8	303	0.40	60.787	173	645	0.37	129.07	543	1286	0.23	257.26
20	74.1	313	0.42	62.623	170	816	0.47	163.28	506	1787	0.35	357.50
30	74.3	307	0.41	61.493	169	846	0.49	169.26	478	2483	0.51	496.64
70	74.2	308	0.41	61.729	168	831	0.49	166.37	459	2542	0.55	508.57

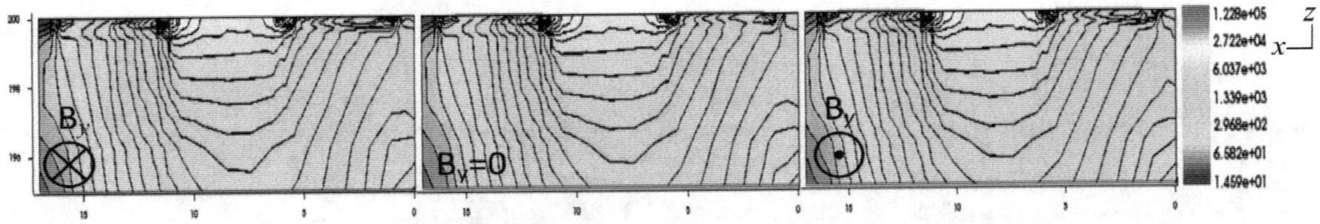

(a) $T = 5$ μm

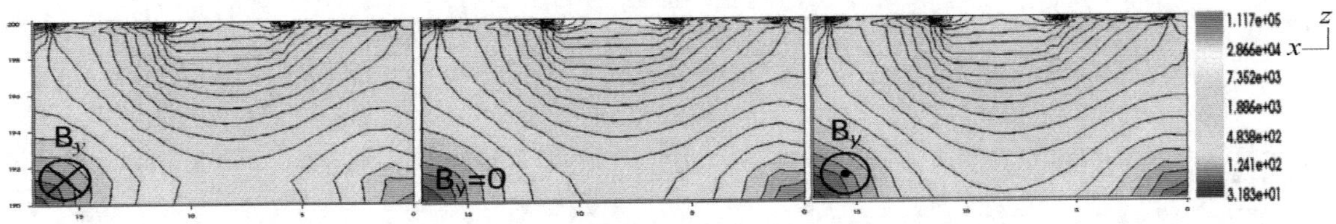

(b) $T = 10$ μm

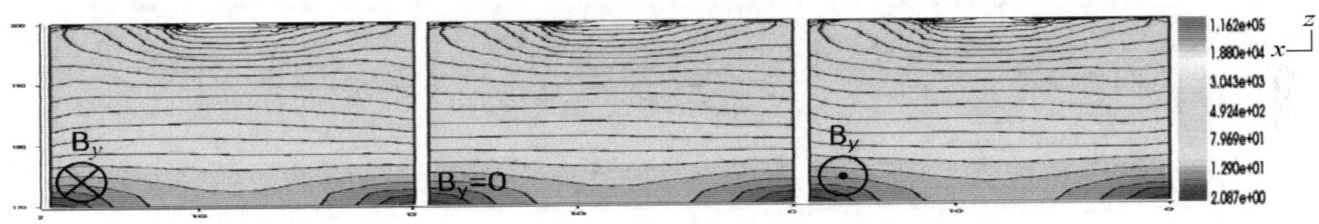

(c) $T = 30$ μm

Figure 4. Current density distribution of resistor $L = 5$ μm.

complete covered by substrate thickness and the longer resistor is also covered by the substrate thickness.

As the results mentioned before, the model for discussion is shown in figure 6. The resistor is biased by constant current will create the current paths between contacts. The current depth t in figure 6 is the depth that corresponds to the biased current value. The current depth can be written as in the term of resistance in table 1

$$R = \frac{\rho L}{A} = \frac{\rho L}{W_d t} \tag{4}$$

$$t = \frac{\rho L}{W_d R} \tag{5}$$

where R is the minimum saturation resistance, ρ is resistivity and W_d is the width of current. The cross section area A is defined as the product of width W_d and depth t. The current density is reduced when the length L of device increases. The current density J can be written as

$$I = \frac{J}{A} = \frac{J}{W_d t} \tag{6}$$

The lower current density J, the longer current depth t is corresponding to constant current I. The resistance of longer substrate thickness will give the lower resistance because the cross section area is increased from the current depth. The substrate thickness T that shorter than current depth t will limit the sensitivity of magnetoresistance. The carrier current distribution is limit by substrate thickness. The substrate thickness that cover the depth of current t, the carrier current can distribute completely

IOP Conf. Series: Materials Science and Engineering **855** (2020) 012008 doi:10.1088/1757-899X/855/1/012008

in bulk that the sensitivity is maximum and will be saturated for the substrate thickness long over the current depth *t*.

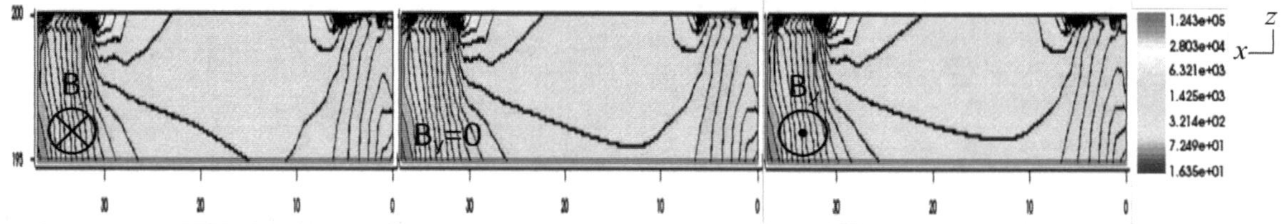

(a) $T = 5$ μm

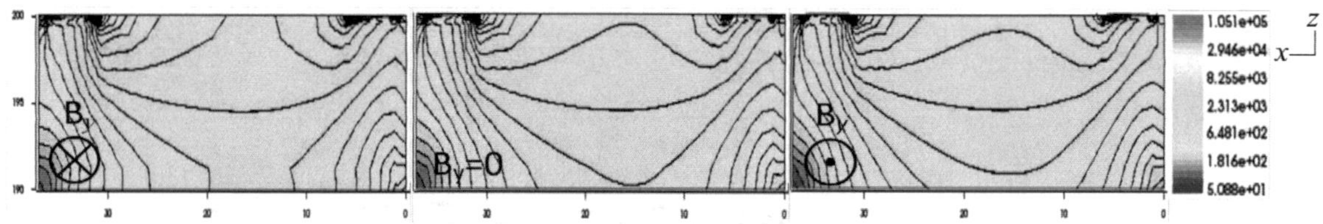

(b) $T = 10$ μm

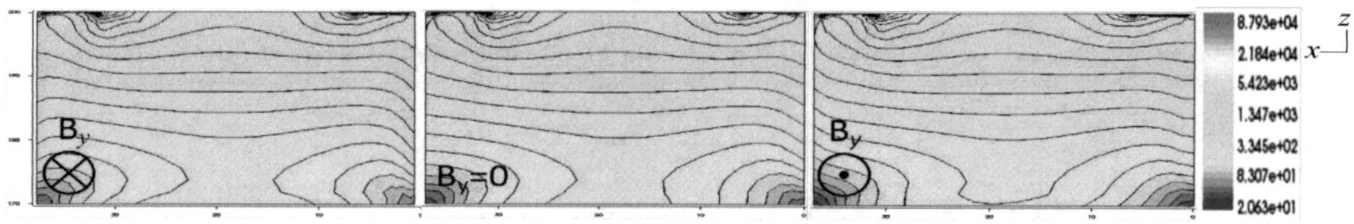

(c) $T = 30$ μm

Figure 5. Current density distribution of resistor $L = 25$ μm.

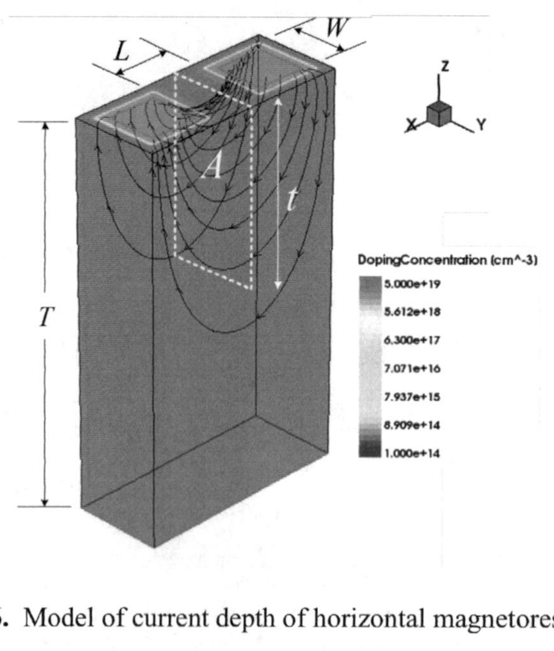

Figure 6. Model of current depth of horizontal magnetoresistance.

5. Conclusion

The substrate thickness affects to horizontal magnetoresistance. The depth of current is the depth that represent to the cross section area corresponding to current. The substrate thickness shorter than current depth will limit current distribution, the conductance and sensitivity. The longer substrate thickness will increase current distribution, conductance, sensitivity and *% MR*. The substrate thickness that longer than current depth will allow the current distribution completely and the resistance, sensitivity and *% MR* are saturated. The longer resistor at the same current will reduced the current density so the depth of current will increase to get the same constant current. The effect of substrate thickness should be considered for operation with horizontal magnetoresistance.

6. References

[1] R. S. Popovic, Hall Effect Devices, 2nd ed., IoP Publishing Ltd., London, 2004.
[2] C. S. Roumenin, "Solid State Magnetic Sensors" Elsevier Science B. V.,Amsterdam, 1994.
[3] E. Ramsden, "Hall-Effect Sensors: Theory and Applications", Elsevier Inc., USA, 2006.
[4] A. M. Huiser and H. P. Baltes, "Numerical modeling of vertical Hall effect devices" IEEE Electron Device Lett., vol. EDL-5, pp. 482-484, 1984.
[5] R. S. Popovic, The Vertical Hall-Effect Device, IEEE Trans. Electron Devices, Vol. EDL-5, No.9, 1984.
[6] T. Phetchakul, S. Poosawat and A. Poyai, "The effect of deviation current to 5-contacts vertical Hall device", ECTI-CON 2016, 28 Jun-1 Jul, Chieng Mai, Thailand.
[7] S. V. Lozanova and C. S, Roumein, "Parallel-Field Silicon Hall Effect Microsensors with Minimal Design Complexity", IEEE Sensors Journal., vol.9, No.7,2009.
[8] Y. Chemthung, T. Phetchakul and A.Poyal "Effect of Horizontal Magnetic Field on Magnetoresistance", ECTI-CON 2018, 18-21 Jul, Chieng Rai, Thailand.
[9] E. Jovanovic, Danijela Pantic, and Dragan Pantic, "Simulation of Vertical Hall Sensor in High-voltage CMOS Technology", [Proceeding of TELSIKS, 1-3 October 2003].
[10] J. Pascal, L. Hebrard, J. B. Kammerer, V. Frick, J. P. Blonde, "First Vertical Hall Device in standard 0.35 μm CMOS Technology", Sensor and Actuators A: Physical, vol. 147, 15 September 2008, pp. 41-46.
[11] S. Li and Y. Fu, "3D TCAD Simulation for Semiconductor Processes, Devices and Optoelectronics", Springer New York, 2012.
[12] C. K. Maiti, "Introduction Technology Computer-Aided Design (TCAD)", Pan Stanford Publishing Pte. Ltd., USA, 2017.
[13] T. Phetchakul, Y. Chemthung, and A. Poyai, "Effect of Aspect Ratio on Horizontal Field Magnetoresistance," in Key Engineering Materials, 2019, pp. 327-334.
[14] TCAD Sentaurus Manual, Synopsys, Version L-2016.03.

Acknowledgments

Authors would like to acknowledge Thai Microelectronics Center (TMEC) for simulation program and Faculty of Engineering, King Mongkut's Institute of Technology Ladkrabang for financial support.

Diamond-like Carbon Thin Film Coating for Application on Heterojunction Solar Cells by ECR-CVD System

Tanawit Srisantirut, Weera Pengchan, Toempong Phetchakul

Faculty of Engineering, King Mongkut's Institute of Technology Ladkrabang, Chalongkrung Road, Ladkrabang, Bangkok, 10520, Thailand

60601070@kmitl.ac.th; weera.pe@kmitl.ac.th

Abstract. Diamond-like Carbon (DLC) film has the ability to change the refractive index value achieved high transmittance for potential applications of antireflection coating. DLC films were deposited on heterojunction solar cells by the ECR-CVD (Electron Cyclotron Resonance Chemical Vapor Deposition) method using Acetylene (C_2H_2) and Argon (Ar) gases. DLC coating affects ITO (Indium tin oxide), causing the electrical and resistivity of heterojunction solar cell vary according to the coating time. The surface and thickness of the DLC films was measured with an atomic force microscope. Illuminated *I-V* characteristic was investigated by photo-*I–V* measurements. The suitable acetylene anti-reflection deposition time at 0.40 minutes. Therefore, heterojunction solar cell was fabricated on a substrate using the optimized DLC films as an anti-reflection coating and it obtained an increase efficiency of about 1%.

1. Introduction

During the decade, diamond-like carbon thin films, known as DLC films are of interest to researchers as they have many properties like diamonds, such as chemical inertness, high mechanical strength, high optical transmittance and electrical resistivity. [1] Found several important applications in the fields of solid-state devices, optics, electricity. [2] DLC films have wide energy bandgaps and adjustable refractive index, therefore applies as protective coatings for IR windows and as anti-reflective coatings for solar cells. [3] DLC film was deposited can be prepared by physical or chemical methods, such as sputtering, plasma enhanced chemical vapor deposition (PECVD) and electron cyclotron resonance chemical vapor deposition. (ECR-CVD) [4] Among these methods, DLC films were deposited by ECR-CVD system, because system uses plasma processing technology that allows the simple and low-temperature coating. In addition to keep the temperature level so high that it does not affect the ITO on the surface of the heterojunction solar cell. In this case hydrogen content in the film has an important effect on the ITO properties. In this paper, the experimental process for film deposition uses an ECR-CVD system. Then we used the DLC films as an anti-reflective coating for scale heterojunction solar cell (1 cm× 1 cm) under optimum conditions. We investigated the effect of the acetylene to ITO on electrical properties of DLC coating. Can be studied compared of efficiencies and *I-V* characteristic of heterojunction solar cell with DLC and without DLC.

2. Experimental

IOP Conf. Series: Materials Science and Engineering **855** (2020) 012009 doi:10.1088/1757-899X/855/1/012009

A set of 24 cells with dimensions of 1 × 1 cm^2 was processed using CZ-grown 380μm thick p-type (100) oriented one side polished silicon wafers with a resistivity of 1–10 Ωcm were used to fabricate the heterojunction solar cell. The heterojunction cell was then fabricated by the successive deposition of the a-SiO:H buffer layer and n-μc-SiO:H as an emitter on the polished side of the silicon wafer. An ITO thin film was deposited by RF magnetron sputtering at a substrate temperature of 200°C, followed by the deposition of aluminium fingers as the emitter contacts to complete the fabrication of the devices. p-a-SiO:H was deposited on the backside prior to the Al back contact to make a good ohmic contact. [5] The top layer has a DLC deposition which will vary with the coating times. Then thin films of combined porous DLC structure were fabricated for anti-reflection coating, showing the entire structure in Figure 1. The films are prepared using low-power electron-cyclotron resonance (ECR) plasma-enhanced chemical vapor deposition. System that uses microwave power, frequency 2.45 GHz, power 2 kw, and uses a magnetic field at a concentration of 875 gausses. The first step of the coating film, substrates were cleaned with plasma and using argon gas to etch on the surface of the object to cause residual dirt to fall out. Then, acetylene gas was introduced into the reaction chamber in order to grow the film at room temperature. The substrate was heated only by the plasma during the growth of the film and the deposited acetylene at 45 cc. Which varies with time in the range of 0.20 to 4 minutes. We changed the coating time in order to investigate on the surfaces and electrical properties of the DLC film. The deposition conditions are shown in Table 1. The surface morphology, thickness of the DLC films under different growth parameters were examined by atomic force microscope. (AFM Workshop TT-2) Analytical bonding structures are obtained from Raman spectroscopy. Once the film has been deposited, the results will be electrically measured from the photo-*I–V* measurements for IV characteristic.

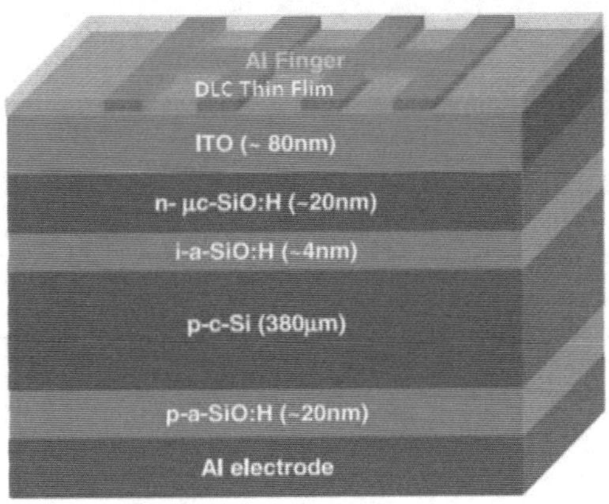

Figure 1. Structure of a heterojunction solar cell coating with DLC

Table 1. Deposition parameters of DLC films

Substrates	Heterojunction solar cell
Etching gas	argon (Ar): 15 cc
Deposition gas	acetylene (C_2H_2): 45 cc
Working pressure	1.5 mTorr
Microwave power	100 W
Deposition time	0.20, 0.40, 1, 2, 3 and 4 min
Substrate temperature	room temperature

3. Results and discussion

In order to determine the change in the surface topography, AFM was used to investigate visualize conformation and morphologic characteristics resulting from deposition of DLC films. The results show that the thickness of the film will change with the deposited time. The scale that is analyzed at 5 x 5 μm. As seen from Figure 2. A significant change in the thickness were observed following the anti-reflection, i.e. for DLC, If the film has a thickness increasing the efficiency of the incident light down to less. This kind of lower roughness of the DLC films is originated from a smoothing effect due to the uniform surface caused by the major argon bombardment.

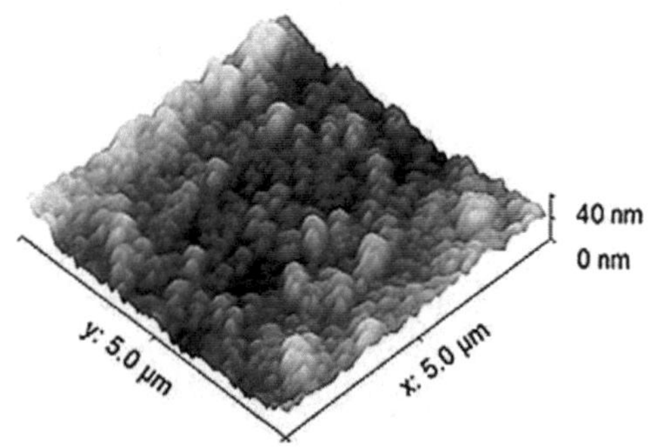

(a)

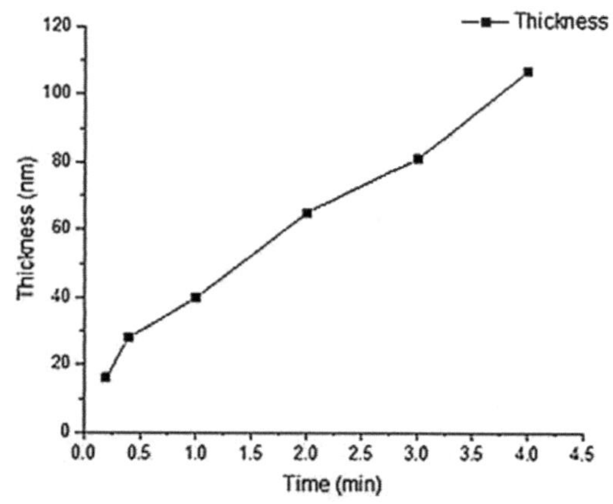

(b)

Figure 2. AFM image of DLC coatings for (a) surface morphology (b) Relationship between thickness and times

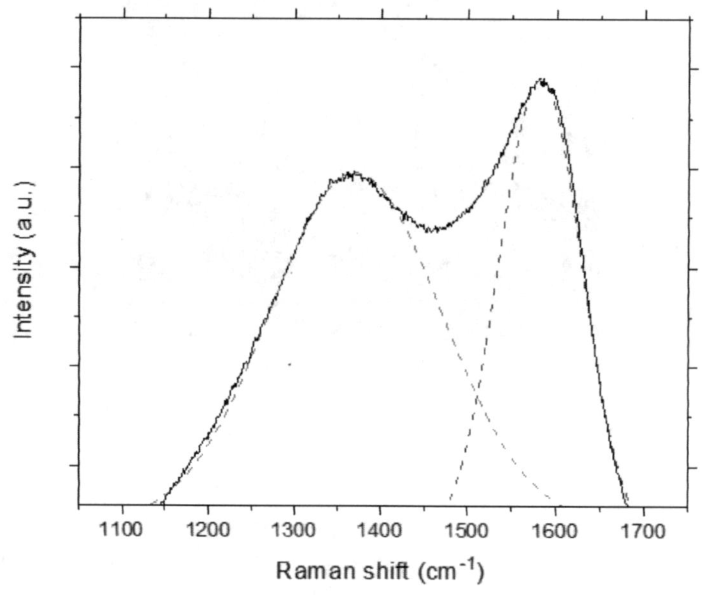

Figure 3. Raman spectra for DLC films deposited

Figure 3 shown the Raman spectra of the films deposited using acetylene to hydrogen at 45 cc. Raman spectra is widely used for studying the structural properties of DLC films, due to its ability to distinguish between the sp^3 and sp^2 bonding. The result of the experiment, Raman peak approximately 1355 cm^{-1} and 1550 cm^{-1}. Which indicates that the film that has been accumulated is a DLC film. (ta-C:H)

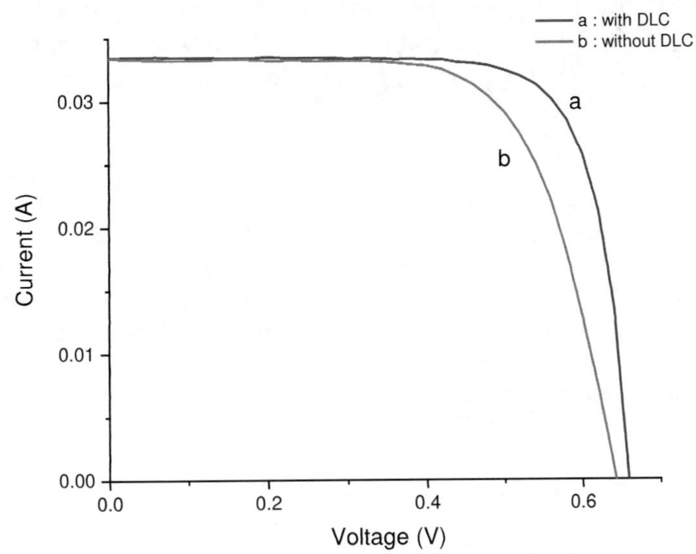

Figure 4. The I-V characteristic of a heterojunction solar cell (a) with DLC and (b) without DLC

There has been research indicating that short-circuit current (I_{sc}) increases as function of DLC until a value of the DLC thickness and then I_{sc} decreases for higher coating thicknesses. The indicates a change in the DLC layer optical properties with preparation conditions. [6] Figure 4 depicts the current-voltage (*I-V* characteristic) at AM 1 illumination for (a) with DLC deposited onto heterojunction solar cell (b) without DLC coating. It can be seen that the short-circuit current slightly increased from 33.33 mA to 33.51 mA.

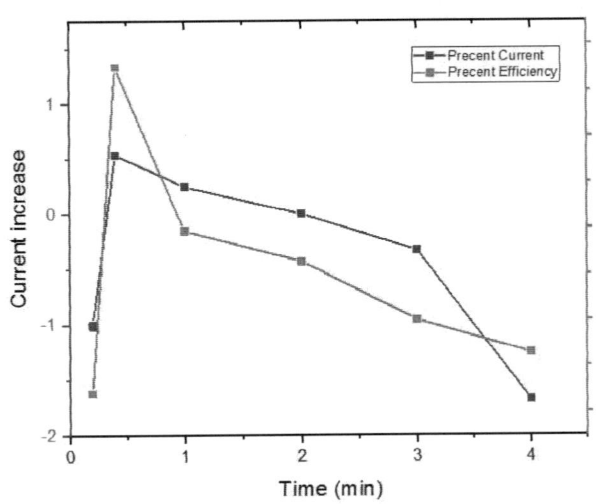

Figure 5. The I-V characteristic of a heterojunction solar cell with DLC and without DLC

From the analysis in Figure 5 of the percent of current and percentage of efficiency compared to the increase in time, it is found that the film that accumulates in the optimal period is 0.40 min. The current and efficiency percentage increase by 0.6 %, 11% as a result of the suitable thickness of film and these active hydrogen species could remove weakly bound oxygen in the depositing films. The improvement of the film conductivity with regard to the annealing treatment under hydrogen atmosphere is explained due to the removal of excess oxygen incorporated and the passivation of the defects in the films [7]. The electrical conductivity depends on the carrier concentration and the mobility, which changes according to the accumulation and therefore affects the resistance.

4. Conclusions

This study confirmed the possibility of using DLC films as a protective anti-reflecting coating for heterojunction solar cells. By examining the effects of acetylene to ITO the electrical properties of the DLC film deposited using the ECR-CVD method. The results of the AFM show that surface morphology and thickness of film. Coating film for a long time causing the thickness to increase, affecting the short circuit current and doping bonds cause changes in substances. The *I-V* characteristic response of the heterojunction solar cell showed an improvement after the deposition of the DLC anti-reflection coating. The efficiency of solar cells increased from 1%. Open-circuit voltage was weakly influenced by the anti-reflection coating

5. References

[1] A. Grill, "Diamond-like carbon: state of the art," Diamond and Related Materials 8, 1999, pp.428- 434.

[2] K. OOHIR, "Characteristics and Applications of DLC film," NTN TECHNICAL REVIEW No.77, 2009 pp.90-95

[3] Won Seok Choi, Kyunghae Kim, Junsin Yi and Byungyou Hong, "Materials Letters 62," 2008, pp577-580

[4] Tanawit Srisantirut and Weera Pengchan, "Key Engineering Materials," Vol. 814, pp47-52

[5] Chandan Banerjee, Jaran Sritharathikhun, Akira Yamada and Makoto Konagai, "Journal of physics 41," 2008, 185107

[6] K.Ait-hamouda, A. Ababou, N. Ga bouze"Materials science forum," Vol. 609, pp179-182

[7] Kurt-Ulrich Ritzau, Torge Behrendt, Daniele Palaferri, Martin Bivour and Martin Hermle, "Thin Solid Films 599," 2016, pp161-165

Acknowledgements

This work is supported by Solar Energy Technology Laboratory (STL)。

Review of Road Materials Self-healing: Problems and Perspective

Sergey Inozemtcev, Evgeniy Korolev

REC "Nanomaterials and Nanotechnology", NRU MGSU, 129337 Yaroslavskoe hw. 26, Russia

inozemtsevss@mail.ru

Abstract. Materials with the ability to restore their own functionality are a promising type of smart materials for road construction. The use of encapsulated functional modifiers is a common solution aimed at implementing self-healing technology. The paper presents general trends in the development of self-healing technology in the field of road materials. A mathematical interpretation of the destruction model of self-healing materials is proposed. The direction of technology development is proposed for the formulation of general requirements and quality indicators of self-healing materials for construction, characterizing the degree of their effectiveness depending on the conditions of use.

1. Introduction

The search for technological solutions to ensure the high reliability of building materials, which lead to an increase in the period of operation of the product or structure, is an actual direction in scientific research. Operating conditions vary over a wide range by type and intensity of factors. Road building materials are subject to the following factors: static or dynamic mechanical stresses, periodicity of climatic conditions (temperature, humidity, ultraviolet, oxygen, etc.), exposure to anthropogenic and natural chemical and biological agents. Usually, new technological solutions reduce the influence of some key factors, since the variability of the superposition of factors is high. The restoration of the performance of products and structures in practice occurs by repair and restoration work, which require economic costs.

For the development of the construction industry, materials with a unique set of properties, more functionality, less material consumption of production are required. A promising solution in the field of materials science to increase the durability of structures is the creation of “smart” materials [1, 2]. The formation of their own internal impacts, which ensures the maintenance of structure parameters at the required level, is a key condition for classifying the material as “smart”. “Reactions” of the material to a change in the structure parameters under the influence of operational factors of natural and / or artificial origin are provided at the material design stage.

For road construction, a promising type of smart materials are materials that have the ability to restore their own functionality in the design provided for by its purpose, which is usually called "self-healing" [3].

The development trend of scientific research in the field of self-healing technology shows (Figure 1) that in this direction there is an annual increase in the number of publications. At the same time, the share of work in the field of road materials by 2018 reached only 3.7%.

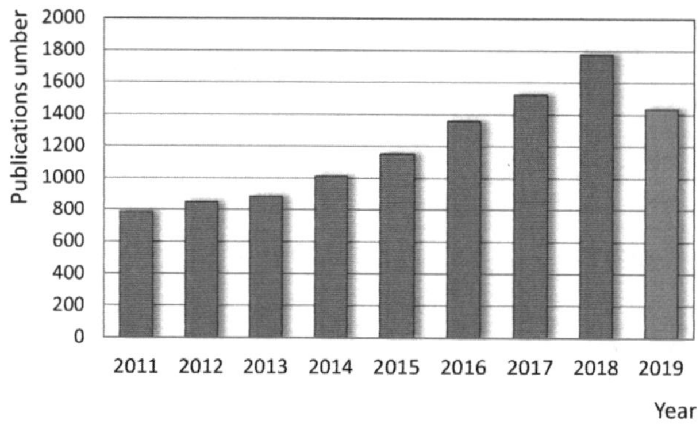

Figure 1. Change in the number of publications in the direction of "self-healing" (according to www.scopus.com, October 2019)

In building materials science, self-healing technologies are used both in polymeric materials and in concrete based on hydraulic or organic binders. The self-healing mechanism is different for each type of building materials [4-9]. Traditional cement concrete, polymer concrete, and asphalt concrete can have their own potential for self-healing.

2. Results and discussion

The number of degradation mechanisms of the material is less than the variability of factors, this allows us to develop new technological solutions to eliminate the consequences of adverse changes in the structure of the material during the operational period. The formation and growth of cracks to critical sizes, after which they begin to spontaneously grow, is the result of the process of accumulation and development of defects (A. A. Griffiths in 1920 [10]). Placing a phase in the bulk of the material that is able to eliminate cracks and not be susceptible to or little susceptible to destruction under the conditions of operation of the material is a natural and fundamental solution. Since there are structural limitations, the reserve of the reducing phase must be designed at the material design stage and localized in areas with intense crack formation.

It is also important to clarify the term "self-healing" material. As a rule, the kinetics of changes in the structurally sensitive material parameter F is described by the differential equation [11]:

$$\frac{dF}{dt} = -k_t F^n,$$

and his solution has the form:

$$\frac{F(t)}{F(0)} = \sqrt[1-n]{1 - k_t \frac{1-n}{F(0)^{1-n}} t}, \quad (1)$$

where $F(0)$ is the value of the structurally sensitive parameter for the initial period of time; n, k_t energy and kinetic constants; t is time.

The kinetics of the change in the value of a structurally sensitive parameter for ordinary and self-healing material is presented in Figure 2.

IOP Conf. Series: Materials Science and Engineering **855** (2020) 012010 doi:10.1088/1757-899X/855/1/012010

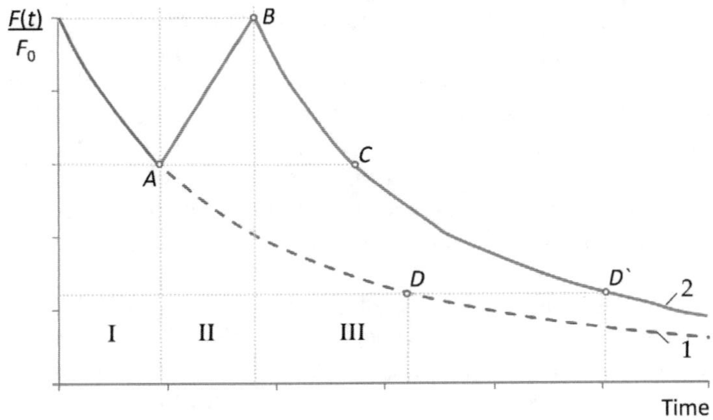

Figure 2. Schematic representation of the kinetics of changes in a structurally sensitive parameter: 1 - ordinary material; 2 - self-healing material

An analysis of equation (1) shows that the rate of achievement of the initial critical state of the material, which is characterized by a decrease in the structurally sensitive parameter throughout the volume of the material and a local violation of its continuity (the formation of a volume defect, for example, micro- and macrocracks, but not magistral cracks), increases with an increase in the energy and kinetic constants. In this case, the durability of the materials is ensured at n and $k_t \to 0$.

The change in the structurally sensitive parameter during operation for ordinary material is described in accordance with formula (1), for which the energy and kinetic constants n_1 and k_{t1} characterize the intensity of the destruction process. The description of this process for a self-healing material is characterized by the presence of three stages of a structurally sensitive parameter change. The change in the structurally sensitive parameter $F_1(t)/F(0)$ can be described using equation (1) in stage I. The self-healing stage of the material (stage II) occurs at the moment of the minimum acceptable value of the structurally sensitive parameter, which is determined during the design of the composition of the material. The rate of the self-healing process should be minimal, therefore, its description can be performed using a linear dependent:

$$F_2(t)/F(0) = a\cdot(t - t_c) + F_1(t)/F(0), \qquad (2)$$

where a is the speed of the self-healing process, t_c is the time after which the critical value $F(t_c)$ is reached and there is a need to restore the material.

The time tc is optimal for the start of the self-healing process, the effectiveness of which is determined by the coefficient a> 0. After the completion of the self-healing stage, the change in the structurally sensitive parameter in stage III is also described by equation (1), for which the self-healing efficiency will be ensured for the values of the energy and kinetic constants $n2 \leq n1$ and $kt2 \leq kt1$.

The effectiveness of the self-healing technology is determined not only by the features of the recovery processes (speed, intensity, structure formation), but also by the features of the destruction of the reduced material, which are characterized by an energy and kinetic constant. Features of the effect on the change in the structurally sensitive parameter of the destruction of the material, due mainly to natural changes in the molecular structure and distribution of structural bonds in the matrix, must be taken during the design process of self-healing material. The intensity of destruction from this factor is reflected in the value of the energy constant n, and the intensity of destructive from mechanical operational influences is reflected in the value of the kinetic constant k_t. In the absence of external mechanical influences $k_t = 0$, even in the presence of internal stresses in the material $n \neq 0$, there will be no additional energy for the formation of defects in the material and the values of the structurally sensitive parameter will not change.

The self-healing process is accompanied by a change in the structure parameters, the formation of new structural elements, the redistribution of structural bonds, and a change in their strength, which should lead to the formation of a structure whose destruction rate is less than the destruction rate of the primary structure of the material. The calculation of increasing the life of the material due to self-healing is evaluated not at the moment of reaching the initial critical state of the material (p. *A* and *C* in Figure 2), but at the moment of reaching the minimum value of $F(t)/F(0)$ or $F2(t)/F(0)$, respectively, i.e., D and $D`$ in Figure 2.

Self-healing in a material extends to a separate category of defects that this process can influence. These include defects with parameters not exceeding the value, the corresponding critical state when the process of their further development acquires an avalanche-like character.

In this regard, a "self-healing" material is a material capable of initiating the process of eliminating structural defects, the flow rate of which exceeds the rate of their spontaneous growth, and the resistance to operational conditions of the formed structure is equal to or greater than the resistance of ordinary material.

Thus, the development of solutions for the implementation of this principle will increase the life of materials, which is an urgent task of materials science, including in the direction of road construction.

Traditionally, cement stone occupies from 10 to 30% of the volume in the composition of concrete, of which 20...30% remain not hydrated (clinker stock). The amount of clinker stock in the volume of concrete depends on the degree of dispersion of Portland cement and the water-cement ratio. During the operation of structures made of such concrete under conditions of natural cracking, water can penetrate into the material, which promotes Portland cement (clinker stock) hydration, and the products of this reaction can fill cracks and other defects, which can be described as a self-healing structure [12].

The difficulty of predicting the access of the required amount of water to start the process of hydration for the formation of reaction products sufficient to stop the defect is the main problem of implementing the technology based on the hydration potential of cement grains inside the concrete.

However, the features of the hydration potential can be used in the production of asphalt concrete using a Portland cement filler. In this case, water penetrating under the bitumen films promotes the start of the hydration process and the formation of hydrosilicate crystals that penetrate the film, increase adhesion and prevent its peeling [13].

Developments using alkaliphilic endospore-forming bacteria [14-20], which starts the process of biomineralization of concrete with subsequent precipitation of calcite [21], are common solutions for creating self-healing material. Bacteria of the genus bacillus are used to provide a regenerative process, which is added to the concrete together with the calcium salt of lactic acid (calcium lactate) and nutritional supplements [22-24]. Bacteria begin their life after interacting with water, consume oxygen, eating calcium lactate, which is subsequently converted to calcium carbonate. Thus, soluble calcium lactate is converted to insoluble calcite, which helps to stop cracks and pores inside the material, preventing moisture from penetrating and increasing frost resistance [25]. At the same time, the corrosion resistance of reinforcing elements in concrete increases when bacteria consume oxygen [26].

Bacterial carriers are used to prevent premature activation of the reproduction process, which concentrate spores and prevent their contact with water during the preparation of the concrete mixture.

In [27, 28], the bacteria carrier is expanded clay aggregate, which is saturated together with a nutrient solution and added to the concrete mix. However, the low strength of the aggregate used leads to a decrease in the compressive strength of concrete to 50%.

An alternative method of encapsulating bacterial spores with nutrients is a hydrogel, which has the ability to absorb and retain moisture [29]. The encapsulation method presented allowed using calcium carbonate to stop cracks up to 0.5 mm wide and reduce water permeability by 68%.

The most widespread methods of encapsulating reducing agents are various capsules, which are made on the basis of polymeric compounds that form capsules during curing [30-34].

The decrease in the physicomechanical properties of concrete is the main problem in the use of capsules with a reducing agent. At the moment, there is not enough data confirming the improvement of physical and mechanical properties of the material after recovery using various techniques for the synthesis of encapsulated modifiers.

The implementation of self-healing technology in asphalt concrete is associated with its structural features and thermoplastic properties. The increase in the strength characteristics of asphalt concrete in accordance with regulatory documents does not ensure its durability in operating conditions, which is associated with the peculiarity of the properties of bitumen [35-37]. An increase in the tensile strength of asphalt concrete at high temperatures leads to an increase in the tensile strength at low temperatures, which negatively affects its fragility.

The balance between maximum strength at positive temperatures and minimum brittleness is one of the conditions for ensuring durability. Durability is the maximum possible lifetime of the asphalt concrete pavement, in which the serviceability is maintained and repair or restoration is not carried out, possibly in the case when the state of the material structure does not change or the ability to respond to external factors without the formation of defects is preserved. Irreversible processes occur leading to the formation of defects and disruption of the structure in asphalt concrete during operation of the pavement under the influence of weather and climate factors and traffic load.

The durability of asphalt concrete is inversely proportional to the speed and intensity of the course of destructive processes. Thus, to increase the service life of the asphalt concrete pavement, solutions are needed to ensure the duration of the state of the structure of asphalt concrete without defects due to giving it unique properties, independently restore the integrity of the composite and its ability to resist influencing factors.

Various types of impregnators are used to extend the service life of asphalt concrete pavement during the operation phase in order to prevent the development of primary defects [38-42]. Impregnating compositions after surface treatment of the asphalt concrete coating diffuses into the material, restoring the molecular composition of the asphalt binder, which leads to a decrease in stiffness [43, 44].

Rejuvenators, low molecular weight organic compounds, are the main component of such impregnating compounds that dissolve in bitumen, compensate for the loss of maltene fractions during aging during operation [45-52].

However, the penetrating ability of anti-aging compositions depends on rheological properties and does not exceed 20 mm [53], therefore, the restoration effect extends only to the surface layers of the coating. Stopping traffic to use asphalt restoration technology using impregnating rejuvenators is also a drawback.

The use of functional “containers” with a rejuvenating agent or modifier, in the form of capsules or hollow fibers, allows you to get rid of these drawbacks. The implementation of this solution will allow asphalt concrete to independently respond to structural changes in order to restore the functional state during operation. Such self-healing material is classified as “smart.”

It should be noted that bitumen has its own potential for self-healing, which depends on both physical and chemical properties [54]. Self-healing of a binder in asphalt concrete occurs in the absence of a dynamic effect, which is aimed at restoring fatigue damage in the material. The degree of recovery is determined by the relaxation time [54-56].

The result of the self-healing process is the restoration of the integrity of the structure of physicochemical bonds by means of wetting the surfaces of cracks, diffusion of molecules between surfaces and arbitrary scattering of molecules, providing hardening [57, 58]. In bitumen, self-healing can occur at the molecular level through reversible hydrogen bonds with the formation of new cross-links and chains [58-60] through ditopic and tritopic molecules [60]. The intensity of the recovery process in bitumen depends on the temperature conditions in which it will be in the period of rest from repeated loads. The recovery process of bitumen proceeds more intensively at a higher temperature. This feature of bitumen is used in the recycling of asphalt concrete.

The self-healing potential of bitumen is used in the technology of induction recovery of asphalt concrete pavement, which was developed at the Delft University of Technology [62]. In accordance with [63-65], 5…7% of the metal fiber is introduced into asphalt concrete, which, using a special induction installation, is exposed to a high-frequency alternating electromagnetic field and is heated into the pavement. Bitumen around heated metal fibers is melted above the softening temperature, recovery occurs and defects in the binder are eliminated [55, 66, 67]. The healing process can be repeated at least five times with an equal degree of recovery (Figure 3).

The temperature gradient across the thickness of the pavement is a drawback of such a technology that occurs during induction heating (Figure 4).

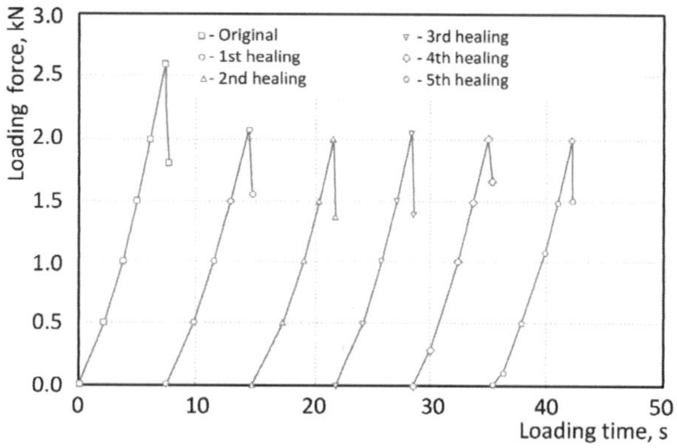

Figure 3. Kinetics of changes in maximum load [64]

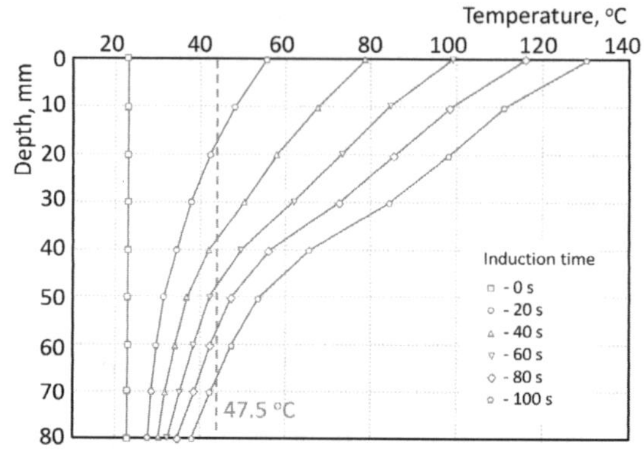

Figure 4. Dependence of the vertical temperature distribution of the duration of induction heating [65]

The temperature difference on the surface and in the volume at a depth of 80 mm can reach more than 90 °C with an increase in the time of induction exposure with a power of 8.3 kW and a frequency of 123 kHz. At the maximum depth, the temperature of the material is optimal for recovery, which is equal to the softening temperature of bitumen – 47.5 ° C, which is not achieved [65].

Thus, the use of induction heating to restore the entire thickness of the pavement is currently impossible. The need to develop special induction equipment increases the cost of introducing technology.

The uniformity of the self-healing process in the volume of the material can be achieved using encapsulated functional agents, which are added to the composition of the asphalt mix together with other components. However, the use of encapsulated modifiers or bacteria in asphalt concrete is associated with a number of technological features of the preparation of the asphalt mix.

Capsules with a reducing agent should be heat-resistant at temperatures of asphalt-concrete mixture preparation - from 140 to 185 ° C and resistant load compaction of the asphalt-concrete mixture.

There are various technologies for the production of microcapsules containing a reducing agent for asphalt concrete, which differ in the initial components, technological process, and product properties.

In work [68], two-stage coacervation technology is used for the manufacture of microcapsules based on a methanol-melamine-formaldehyde-modified methanol prepolymer. The technology of using melamine formaldehyde in the manufacture of capsules with an anti-aging agent is environmentally unsafe [69].

In [70], the options for obtaining the capsule shell from cellulose and polymer for an emulsified reducing agent in water were considered.

The authors of [71] proposed a three-stage method for producing microcapsules, in which at the first stage styrene-maleic anhydride and water are mixed for 2 hours at a temperature of 50 °C, then NaOH is added to pH 10 and a reducing agent is added. In the second stage, the resulting emulsion and methanol-melamine-formaldehyde are mixed while heating to 80 °C with a rate of 2 °C/min. In the third stage, the emulsion is kept at 80 ° C for 2 hours, after which the resulting capsules are filtered and dried in a vacuum oven.

In [72], a multistage capsulation technology of a reducing agent was proposed, in which the reducing agent emulsified in water is coated with a urea and formaldehyde coating.

The most technologically simple way to obtain microcapsules was proposed by the authors [73]. Sodium alginate ($C_6H_7O_6Na$), which is the sodium salt of alginic acid, is used to make calcium alginate capsules. Sodium alginate is poorly wetted by water, individual particles of alginate powder are actively aggregated in water with the formation of agglomerates. Sodium alginate is added to water and mixed for 2 minutes using an overhead mixer with a drive rotation speed of at least 2000 rpm [73]. The destruction of agglomerates occurs in the process by high-speed mixing and a colloidal solution is formed. A reducing agent is added to the resulting suspension and mixed for 2 minutes. The resulting emulsion is divided into capsules through a dropping funnel with an outlet of 0.71 cm. The division is

IOP Conf. Series: Materials Science and Engineering **855** (2020) 012010 doi:10.1088/1757-899X/855/1/012010

carried out in an aqueous solution of calcium chloride. After filtering and washing, the obtained capsules were dried in a dry oven at a temperature of 40 °C for 12 hours.
A suspension of sodium alginate is added dropwise to the calcium chloride solution, where gelation of the alginate occurs (at pH<4) in the next stage of capsule preparation. The mechanism of the formation of alginate gels consists of joint binding of calcium ions between macromolecule ribbons located in one line, which have pores or cavities of 0.19...0.20 nm in size corresponding to the diameter of Ca^{+2} ion [74]. Intensive gelation occurs when pores are filled with calcium ions; crosslinking of macromolecules through substituted sodium cations Na^{+} occurs. The separated drops of an alginate gel with an emulsified reducing agent are dried after washing with distilled water. Evaporation of excess moisture from the gel and a decrease in volume occurs during the drying process, the emulsion breaks down inside an individual alginate gel particle, calcium-alginate capsules with a reducing agent are formed.
Rejuvenators, organic oils with different molecular weights are used as a reducing agent by the authors of all the works reviewed. The mechanism of action of such a reducing agent in asphalt concrete is the dissolution of bitumen components aged during operation and local reduction of fragility.
The result of encapsulation technologies is the production of micro containers containing a reducing agent, which differ in the initial components and properties of the resulting capsules. The main characteristics of such containers with various shapes are presented in table 1.
Capsules obtained by various methods have sizes from 10 μm to 3 mm, in which the encapsulated reducing agent is predominantly vegetable oils [68, 69, 72-80]. Industrial rejuvenator is an Alternative variant of the reducing agent, which is a mixture of low molecular weight compounds and oils [70-72, 75].
As a rule, the capsule shape is spherical or ellipsoidal, which facilitates their use in the process of mixing with other components of the asphalt mix. The technology has a significant drawback if the container for the reducing agent is fiber: premature destruction of the containers during mixing and during operation.
The highest content of the reducing agent in the capsules is achieved in the shells of calcium alginate, which is explained by the simplicity of the preparation technology and a wide range of variable dosages of the components.
Capsules produced by various technologies must have a strength that ensures their integrity during the preparation of the asphalt mix and its compaction.
The quality of the self-healing technology depends on the technological properties of the capsules and the restoring properties of the encapsulated agent. A single methodology for controlling the ability of a material to self-healing does not exist at present. There are no criteria for assessing the ability of a material to independently respond to external conditions and eliminate an adverse effect on the properties or structure of the material.
In [81], quality criteria were proposed that reflect the effectiveness of a material with self-healing ability: the degree of restoration of the state of the structure; rate of restoration of the state of the structure; the durability of the restored structure and the timeliness of initiating the recovery process. However, a large number of empirical studies are necessary to select property indicators that characterize each of the quality criteria.
Quality indicators, which reflect the self-healing ability of asphalt concrete, are presented in table 2.

Table 1. The main properties of capsules containing a reducing agent

№	Capsule material	Form	Size	Reducing agent (RA)	Volume RA	Reference
1	Methanol-melamine-formaldehyde prepolymer	Sphere	Diameter – 100 μm	Sunflower oil	–	[68, 69]

2	Sepiolite	Sphere	–	Industrial rejuvenator	40 %	[70]
3	Polymer	Sphere	Diameter – 1.00…1.30 мм	Industrial rejuvenator	80 %	[70]
4	Methanol-melamine-formaldehyde prepolymer	Sphere	Diameter – 10…20 мкм	Industrial rejuvenator	50 %	[70]
5	Methanol-melamine-formaldehyde prepolymer	Sphere	Diameter – 57.5 мкм	Oil containing polar epoxy groups	80 %	[72]
6	Calcium alginate	Sphere	Diameter – 2.90 мм	Sunflower oil	48 %	[73]
7	Calcium alginate	Sphere	Diameter – 1.95 мм	A mixture of rejuvenator with an olefin-maleic anhydride polymer solution	56 %	[75]
8	Calcium alginate	Sphere	Diameter – 2.90 мм	Sunflower oil	46 %	[76]
9	Calcium alginate	Cylinder	Diameter – 0.5 мм; Length – 1.5 мм	Sunflower oil	70 %	[77, 78]
10	Calcium alginate	Ellipsoid	Diameter – 0.98…1.74 мм	Sunflower oil	60 %	[79]
11	Calcium alginate	Sphere	Diameter – 1.95…2.45 мм	Sunflower oil	90 %	[79]
12	Calcium alginate	Sphere	Diameter – 2.5 мм	Sunflower oil	94 %	[80]

Table 2. Self-healing performance indicators of asphalt concrete

№	Indicator	Condition for healing		Effect	Reducing agent	Reference
		Time, h	Temperature, °C			
1		4	20±2	73…90 %	A mixture of a reunitor with a polymer solution of olefin-maleic anhydride	[75, 79]
2	3-point bending strength	6…192	20±2	up to 53 %	Sunflower oil	[76]
3		3	20±2	50 %	Rejuvenator Modiseal R20	[78]
4		3	20±2	30 %	Sunflower oil	[78]
5		3	20±2	24 %	Waste sunflower oil	[78]
6		6…216	20±2	54…95 %	Sunflower oil	[82]
7		5…216	20±2	20…55 %	Sunflower oil	[84]
8	Tensility	4	21±1	15…90 %	Nanoparticles	[83]
9	Stiffness	0.5…1.5	21±1	80…100 %	Nanoparticles	[83]
10	Tensile strength	6…48	21±1	up to 80 %	Nanoparticles	[83]
11	Crack resistance	20	23±2	14…19 %	Rejuvenator Modiseal R20	[85]
12		20	20±2	–	A mixture of a reunitor with a polymer solution of olefin-maleic anhydride	[86]

13	Fatigue life	20	20±2	–	A mixture of a reunitor with a polymer solution of olefin-maleic anhydride	[86]
14	Bending strength	72	30±2	50…89 %	Rejuvenator	[87]

All proposed methods for assessing the self-healing ability of materials are reduced to calculating the coefficient of relative change in the measured indicator:

$$HL = \frac{X_h}{X_0}$$

where X_0 and X_h are the index of material properties before and after self-healing, respectively.
In accordance with [81], this coefficient can be assigned to the category of indicators that reflect the degree of restoration of the state of the structure, which is insufficient for an objective assessment of technological solutions.
Fatigue life, which reflects the dynamic loading pattern in the pavement, is used as an indicator for assessing the self-healing effect of the material only in [86]. In other cases [75, 76, 78, 82-85, 87], the authors use property indicators characterizing the operational state of the material at the time of failure, when the defectiveness of the structure is ultimate — the structure is destroyed. However, in the pavement, the initiation of a self-healing process should occur at the stage preceding the destruction of the structure.
There is no unified evaluation system, which allows you to objectively compare the effectiveness of microcapsules with a reducing agent produced by various technologies and evaluate the effects of prescription and technological factors on the self-healing process and its result.
Thus, it is necessary to improve the technological aspects to obtain effective capsules and develop methods for evaluating the effectiveness of technologies: the degree of change in the defectiveness of the structure; the rate of its change during recovery; the kinetics of the formation of defects in the structure after the self-healing process. A key task in the field of self-healing technology is to establish the defectiveness of the material at which the start of the self-healing process is the most effective.

3. Conclusion

An analysis of the scientific and technical literature shows a growing interest on the part of researchers and technologists in the field of technology of self-healing materials.
Self-healing materials are materials that have the ability to independently initiate processes of restoration of structural parameters without an additional energy cost, the rate of which exceeds the rate of formation of defects.
Existing technical solutions allow you to get capsules containing a reducing agent up to 90...94% of the total volume.
It is necessary to formulate general requirements and quality indicators for self-healing materials characterizing the degree of their effectiveness depending on the conditions of use.

4. References

[1] Bengisu M., Ferrara M. Materials that Move: Smart Materials, Intelligent Design. 2018. PoliMI SpringerBriefs. 139 p. DOI: doi.org/10.1007/978-3-319-76889-2_5.

[2] Shahinpoor M., Schneider H.-J. Intelligent Materials. 2007. 532 p. DOI: doi.org/10.1039/9781847558008.

[3] Ghosh Swapan Kumar. *Self-healing materials: fundamentals, design Strategies, and applications* (1st ed.). Weinheim: Wiley. 2009. VCH. 307 p.

[4] Schlangen E., Sangadji, S. Addressing infrastructure durability and sustainability by self healing mechanisms - Recent advances in self healing concrete and asphalt. *Procedia Engineering*. 2013. Vol. 54. Pp. 39-57.

[5] Zhuang X., Zhou S. The prediction of self-healing capacity of bacteria-based concrete using machine learning approaches. 2019. *Computers, Materials and Continua*. Vol. 59 Issue 1. Pp. 57-77.

[6] Wiktor V., Jonkers H.M. Bacteria-based concrete: From concept to market. *Smart Materials and Structures*. 2016. Vol. 25 Issue 8. 084006.

[7] De Belie N., Wang J. Bacteria-based repair and self-healing of concrete. *Journal of Sustainable Cement-Based Materials*. 2015. Vol. 5. Issue 1. Pp. 35-56.

[8] Nallanathel M., Needhidasan S., Ananya M. Self healing concrete - Facts and potentials. *International Journal of Applied Engineering Research*. 2015. Vol. 10. Special Issue 33. Pp. 24243-24246.

[9] Li W., Dong B., Yang Z., Xu J., Chen Q., Li H., Xing F., Jiang Z. Recent Advances in Intrinsic Self-Healing Cementitious Materials. *Advanced Materials*. 2018. Vol. 30. Issue 17. 1705679.

[10] Griffith A.A. The phenomenon of rupture and flow in solids. *Philosophical Transactions of the Royal Society of London*. 1920. A 221. Pp. 163–198.

[11] Korolev E.V., Bazhenov Y.M., Albakasov A.I. *Radiatsionno-zaschitnye i himicheski stoykie stroitelnye materialy* [Radiation-protective and chemically stable sulfur construction materials]. Penza, Orenburg: IPK OGU. 2010. 364 p. (In Russian).

[12] Breugel K. Self healing concepts in Civil Engineering for sustainable solutions: Potential and constraints. Proceedings of the Second International Conference on Self-healing Materials, Berlin. 2009.

[13] Inozemtcev S.S., Korolev E.V. Interaction process on the phases interface "bitumen – dispersed phase from cement stone". *Magazine of Civil Engineering*. Vol. 82. Issue 6. Pp. 60-67.

[14] Jonkers H.M. Self healing concrete: a b iological approach. In Self healing materials - An alternative approach to 20 centuries of materials science. 2007. Pp. 195 –204

[15] Jonkers H.M., Schlange E. Development of a bacteria - based self healing c oncrete. In Tailor made concrete structures - new solutions for our society. Proc. Int. FIB symposium. 2008. Pp. 425 -430.

[16] Jonkers H.M., Schlangen E. Bacteria-based self-healing concrete. *International journal of restoration of buildings and monuments*. 2009. Vol. 15. Issue 4. Pp. 255-265.

[17] Jonkers H.M., Thijssen A., Muijzer G., Copuroglu O., Schlangen E. Application of bacteria as self-healing agent for the development of sustainable concrete. *Ecological engineering*. 2010. Vol. 36 Issue 2. Pp. 230-235.

[18] Zemskov S., Jonkers H.M., Vermolen F.J. An analytical model for the probability characteristics of a crack hitting an encapsulated self-healing agent in concrete. In V.P. Gerdt, W. Koepf, E.W. Mayr & E.V. Vorozhtsov (Eds.), *Computer Algebra in Scientific Computing*. 2010. Vol. 6244. Pp. 280-292.

[19] Jonkers H.M., Loosdrecht M. Biogeocivil engineering. *Ecological engineering*. 2010. Vol. 36 Issue 2. Pp. 97-98.

[20] Zemskov S.V, Jonkers H.M., Vermolen F.J. Two analytical models for the probability characteristics of a crack hitting encapsulated particles: Application to self-healing materials, *Journal of Computational Materials Science*. 2011. Vol. 50. Pp. 3323-3333.

[21] Srinivasa Reddy V., SeshagiriRao M.V., Sushma S. Feasibility Study on Bacterial Concrete as an innovative self crack healing system. *International Journal of Modern Trends in Engineering and Research*. 2015. Vol. 2. Issue 7. Pp. 642-647.

[22] Bansal S., Tamang R.K., Bansal P., Bhurtel P. Biological Methods to Achieve Self-healing in Concrete. *Lecture Notes in Civil Engineering*. 2020. Vol. 38. Pp. 63-71.

[23] Joshi K.A., Kumthekar M.B. and Ghodake V.P. Bacillus Subtilis Bacteria Impregnation in Concrete for Enhancement in Compressive Strength. *International Research Journal of Engineering and Technology* (IRJET). 2016. Vol. 3 Issue 5. Pp. 1229–1234.

[24] Sarkar B.C., Sarkar M. Genetically modified Bacillus subtilis bacterial strain for self-healing and sustainable green bio-concrete. *Green Chemistry*. 2015. Vol. 5. Issue 3. p. 4172.

[25] Depaa RA. B. and Felix Kala T. Experimental Investigation of Self Healing Behavior of Concrete using Silica Fume and GGBFS as Mineral Admixtures. 2015. *Indian Journal of Science and Technology*. Vol. 8. Issue 36. DOI: 10.17485/ijst/2015/v8i36/87644.

[26] Sakina Najmuddin Saifee, Divya Maheshbhai Lad, Jayesh Rameshbhai Juremalani. Critical appraisal on Bacterial Concrete. IJRDO-*Journal Of Mechanical And Civil Engineering*. 2015. Vol. 1. Issue 3. Pp. 10-14.

[27] Jonkers H.M., Thijssen A., Muyzer G., Copuroglu O. Schlangen E. Application of bacteria as self-healing agent for the development of sustainable concrete. *Ecological Engineering*. 2010. Vol. 36. Issue 2. Pp. 230-235.

[28] Jonkers H.M., Schlangen E. Development of a bacteria-based self healing concrete. In Tailor made concrete structures - new solutions for our society. Proc. Int. FIB symposium. 2018. Pp. 425-430.

[29] Wang J., Snoeck D., Van Vlierberghe S., Verstraete W., De Belie N. Application of hydrogel encapsulated carbonate precipitatingbacteria for approaching a realistic self-healing in concrete. *Construction Building Materials*. 2014. Vol. 68, Pp.110–119.

[30] Feng X. Self-healing mechanism of a novel cementitious composite using microcapsules. Proc. Int. Conf. on Durability of Concrete Structures. 2008. Pp. 195-204.

[31] Mihashi H., Kaneko Y., Nishiwaki T., Otsuka K. Fundamental Study on Development of Intelligent Concrete Characterized by Self-Healing Capability for Strength. *Trans. Japan Concrete*. Inst..2000. Vol. 11. Issue 2. Pp. 21-28.

[32] Van Tittelboom K., Wang J., Araújo M., Snoeck D., Gruyaert E., Debbaut B., Derluyn H., Cnudde V., Tsangouri E., Van Hemelrijck D., De Belie N. Comparison of different approaches for self-healing concrete in a large-scale lab test. *Constr. Build. Mater*. 2016. Vol. 107. Pp. 125–137.

[33] Thao T. D. P., Johnson T. J. S., Tong Q. S., Dai P. S. Implementation of self-healing in concrete-proof of concept. IES J.Part A: *Civ. Struct. Eng*. Vol. 2. Issue 2. Pp. 116–125.

[34] Han S., Choi E.K., Park W., Yi C., Chung N. Effectiveness of expanded clay as a bacteria carrier for self-healing concrete. *Applied Biological Chemistry*. 2019. Vol. 62. Issue 1. Pp. 19-24.

[35] Inozemtcev S.S., Korolev E.V. Agressivnost' ekspluatatsionnykh usloviy dorozhno-klimaticheskikh zon Rossii [The aggressiveness of operational conditions of road-climatic zones of Russia]. *Nauka i tekhnika v dorozhnoy otrasli* [Science and technology in the road industry]. 2019. Vol. 4. Pp. 24-28. (In Russian).

[36] Inozemtcev S.S., Korolev E.V. Increasing the weathering resistance of asphalt by nanomodification. *Materials Science Forum*. 2019. Vol. 945. Pp. 147-157.

[37] Inozemtcev S.S., Korolev E.V. Ekspluatatsionnyye svoystva nanomodifitsirovannykh shchebenochno-mastichnykh asfal'tobetonov [The operational properties of nanomodified stone-mastic asphalt concrete]. *Vestnik MGSU* [Bulletin of MGSU]. 2015. Vol. 3. Pp. 29-39. (In Russian).

[38] Vysotskaya M.A., Vlasova Ye.A., Kuznetsov D.A., Kurlykina A.V., Shekhovtsova S.Yu. Obzor sostoyaniya segmenta propitochnykh materialov dlya pokrytiy avtomobil'nykh dorog [An overview of the condition of the segment of impregnating materials for road coatings]. *Vestnik BGTU V.G. Shukhova*. [Bulletin of BSTU V.G. Shukhov]. 2018. Vol. 8. Pp. 6-12. (In Russian).

[39] Merkulov V.V., Ibatov M.K., Izmailova G.G., Zhaksybayeva G.SH., Mantler S.N. Primeneniye modifitsirovannoy koksokhimicheskoy smoly dlya obrabotki asfal'tobetonnykh dorozhnykh pokrytiy [The use of modified coke resin for the processing of asphalt concrete pavements]. *Sovremennyye naukoyemkiye tekhnologii* [Modern high technology]. 2016. Vol. 10 Issue 2. Pp. 267-270. (In Russian).

[40] Musa S.S., Rudakova V.V. Perspektivy primeneniya dorozhnykh bitumnykh emul'siy [Prospects for the use of road bitumen emulsions]. *Avtomobil'. Doroga. Infrasturktura* [Car. Road. Infrastructure]. Vol. 3. Issue 5. 2015. Pp. 17-26. (In Russian).

[41] Belyayev N.N., Panevin N.I. Novaya rol' propitok [The new role of impregnation]. *Avtomobil'nyye dorogi* [Car roads]. Vol. 11. 2016. Pp. 82-86. (In Russian).
[42] Shekhovtsova S.Yu., Korolev E.V. Obzor sovremennogo opyta ispol'zovaniya reyunivatorov dlya reversinga asfal'tobetonnykh pokrytiy [A review of current experience with the use of reunites for reversing asphalt concrete pavements]. *Regional'naya arkhitektura i stroitel'stvo* [Regional architecture and construction]. Vol. 3. Issue 36. Pp. 5-16.
[43] Brownbridge J. The Role of an Asphalt Rejuvenator in Pavement Preservation: Use and Need for Asphalt Rejuvenation. Compendium of Papers from the First International Conference on Pavement Preservation. Newport Beach CA, USA. 2010. Pp. 351-364.
[44] Zaumanis M., Mallick R.B., Poulikakos L., Frank R. Influence of six rejuvenators on the performance properties of Reclaimed Asphalt Pavement (RAP) binder and 100% recycled asphalt mixtures. *Construction and Building Materials*. 2014. Vol. 71. Pp. 538-550.
[45] P. Cong, J. Wang, K. Li, S. Chen Physical and rheological properties of asphalt binders containing various antiaging agents. *Fuel*. 2012. Vol. 97. Pp. 678-684.
[46] Bergendahl J. Environmental Effects of Coal Tar-Based Pavement Rejuvenators. 2017. 76 p.
[47] Zhang, Yu., Molenaar A.A., Shaopeng W. Assessment of effectiveness of rejuvenator on artificially aged porous asphalt concrete. *Construction and Building Materials*. 2016. Vol. 110. Pp. 286-292.
[48] Mazzonia G., Boccib E. Canestraria F. Influence of rejuvenators on bitumen ageing in hot recycled asphalt mixtures. *Journal of Traffic and Transportation Engineering* (English Edition). 2018. Vol. 5. Issue 3. Pp. 157-168.
[49] Xu G., Wang H., Sun W. Molecular dynamics study of rejuvenator effect on RAP binder: Diffusion behavior and molecular structure. *Construction and Building Materials*. 2018. Vol. 158. Pp. 1046-1054.
[50] Moghaddam T.B., Baaj H. The use of rejuvenating agents in production of recycled hot mix asphalt: a systematic review. *Construction and Building Materials*. 2016. Vol. 114. Pp. 805-816.
[51] Ongel A., Hugener M. Impact of rejuvenators on aging properties of bitumen. *Construction and Building Materials*. 2015. Vol. 94. Pp. 467-474.
[52] Yu X.K., Zaumanisa M., Santos S.D., Poulikakos L.D. Rheological, microscopic, and chemical characterization of the rejuvenating effect on asphalt binders. *Fuel*. 2014. Vol. 135. Pp. 162-171.
[53] Shen J., Amirkhanian S., Miller J. A. Effect of rejuvenating agents on superpave mixtures containing reclaimed asphalt pavement. *Journal of Materials in Civil Engineering*. 2007 Vol. 19. Issue 5. Pp. 376-384.
[54] Bhasin A., Palvadi S., Little D. Influence of Aging and Temperature on Intrinsic Healing of Asphalt Binders. *Transportation Research Record Journal of the Transportation Research Board*. 2011. Vol. 2207. Issue 1. Pp. 70-78.
[55] Tang J., Liu Q., Wu Sh., Ye Q., Sun Y., Schlangenc E. Investigation of the optimal self-healing temperatures and healing time of asphalt binders. *Construction and Building Materials*. 2016. Vol. 113. Pp. 1029-1033.
[56] Xie W., Castorena C., Wang Ch., Kim Y.R. A framework to characterize the healing potential of asphalt binder using the linear amplitude sweep test. *Construction and Building Materials*. 2017. Vol. 154. Pp. 771-779.
[57] Amir Tabaković, Erik Schlangen. Self-Healing Technology for Asphalt Pavements. *Advances in Polymer Science* book series. 2015. Vol. 273. Pp. 285-306.
[58] Fisher H. Self repairing materials – dream or reality? *Natural Science*. 2010. Vol. 2. No. 8. Pp. 873-901.
[59] Qui J., M.F.C. van de Ven, Wu S., Molenaar A.A.A. Investigation of self healing capability of bituminous binders. *Road Materials Pavement Design*. 2009. Vol. 10. Issue 1. Pp. 81-94.
[60] Phillips M.C. Multi-step models for fatigue and healing, and binder properties involved in healing. In: Eurobitume workshop on performance related properties for bituminous binders, 1998, Luxembourg. Paper no. 115.

[61] Cordier P. et al. Self-healing and thermoreversible rubber from supramolecular assembly. *Nature*. 2008. Vol. 451. Pp. 977-980.
[62] Liu Q, Schlangen E, van de Ven M., García A. Induction heating of electrically conductive porous asphalt concrete. *Construction and Building Materials*. 2010. Vol. 24. Pp.1207-1213.
[63] Liu, Q., Schlangen, E., García, Á., Van De Ven, M. Healing of porous asphalt concrete via induction heating. *Road Materials and Pavement Design*. 2010. Vol. 11. Pp. 527-542.
[64] Liu Q., Schlangen E., Van De Ven M. Induction healing of porous asphalt concrete beams on an elastic foundation. *Journal of Materials in Civil Engineering*. 2012. Vol. 25. Pp. 880-885.
[65] Xu Sh., García A., Su J., Liu Q., Tabaković A., Schlangen E. Self-Healing Asphalt Review: From Idea to Practice. *Advanced materials interface*. 2018. Vol. 5. Issue 17. 1800536.
[66] Garcia A. Self-healing of open cracks in asphalt mastic. *Fuel*. 2012. Vol. 93. Pp. 264-272.
[67] Sarsam S., Barakhas S.A. Influence of Load Repetitions and Heating on Micro Crack Healing of Asphalt Stabilized Subgrade Soil. *International Journal Materials Chemistry and Physics*. 2015. Vol. 1. Issue 3. Pp. 399-405.
[68] Su J.F., Schlangen E., Qiu J. Design and construction of microcapsules containing rejuvenator for asphalt. *Powder technology*. 2004. Vol. 235. Pp. 563-571.
[69] Su J.F., Wang Y.Y., Han N.X., Yang P., Han S. Experimental investigation and mechanism analysis of novel multi-self-healing behaviors of bitumen using microcapsules containing rejuvenator. *Construction Building Material*. 2016. Vol. 106. Pp. 317-329.
[70] Raquel Casado Barrasaa, Víctor Blanco Lópeza, Carlos Martín-Portugués Montoliua, Verónica Contreras Ibáñezb, Josefina Pedrajasc, Julio Santarénd Addressing durability of asphalt concrete by self-healing mechanism. Procedia - Social and Behavioral Sciences. 2014. Vol. 162. Pp. 188 – 197.
[71] Su J.F., Schlangen E. Synthesis and physicochemical properties of novel high compact microcapsules containing rejuvenator applied in asphalt. *Chemical Engineering Journal*. 2012. Vol. 198-199. Pp. 289-300.
[72] Bin Xue, Huifeng Wang, Jianzhong Pei, Rui Li, Jiupeng Zhang, Zepeng Fan Study on self-healing microcapsule containing rejuvenator for asphalt. *Construction and Building Materials*. 2007. Vol. 135. Pp. 641–649.
[73] Al-Mansooria T., Micaeloabc R., Artamendid I., Norambuena-Contrerasae J., Garcia A. Microcapsules for self-healing of asphalt mixture without compromising mechanical performance. *Construction and Building Materials*. 2017. Vol. 155. Pp. 1091-1100.
[74] Koryachkina S.Ya., Prigarina O.M. *Nauchnyye osnovy proizvodstva produktov pitaniya: uchebnoye posobiye dlya vysshego professional'nogo obrazovaniya* [The scientific basis of food production: a textbook for higher professional education] / Orel: FGBOU VPO «Gosuniversitet UNPK». 2011. 377 p.
[75] Xua S., Tabaković A., Liua X., Schlangena E. Calcium alginate capsules encapsulating rejuvenator as healing system for asphalt mastic. *Construction and Building Materials*. 2018. Vol. 169. Pp. 379-387.
[76] Al-Mansoori T., Norambuena-Contreras J., Garcia A. Effect of capsule addition and healing temperature on the self-healing potential of asphalt mixtures. *Materials and Structures*. 2018. Pp. 51-53.
[77] Prajer M., Wu X., Garcia S.J., Van der Zwaag S. Direct and indirect observation of multiple local healing events in successively loaded fibre reinforced polymer model composites using healing agent-filled compartmented fibres. *Composites Science and Technology. Interfaces*. 2014. Vol. 6. Issue 5. Pp. 3661–3670.
[78] Tabakovic A., Dirk B, Van Gerwen M., Copuroglu O., Post W., Garcia S.J., Schlangen E. The compartmented alginate fibres optimisationfor bitumen rejuvenator encapsulation. *Journal of Traffic and Transportation Engineering*. 2017. Vol. 4. Issue 4. Pp. 347-359.
[79] Xu S., Tabakovic A., Liu X., Palin D., Schlangen E. Optimization of the Calcium Alginate Capsules for Self-Healing Asphalt. *Applied Sciences*. 2019. Vol. 9 Issue 3. Pp. 468-480.

IOP Conf. Series: Materials Science and Engineering **855** (2020) 012010 doi:10.1088/1757-899X/855/1/012010

[80] Inozemtsev S.S., Korolev E.V. Technological features of production calcium-alginate microcapsules for self-healing asphalt. MATEC Web of Conferences electronic collection. 2018. P. 01008.
[81] Inozemtcev S., Korolev E. Indicators of the effectiveness of self-healing asphalt concrete. XXII International Scientific Conference "Construction the Formation of Living Environment" (FORM-2019) Electronic edition. Ser. "E3S Web of Conferences" 2019. P. 02007.
[82] Al-Mansoori T., Norambuena-Contreras J., Micaelo R., Garciaa A. Self-healing of asphalt mastic by the action of polymeric capsules containing rejuvenators. *Construction and Building Materials*. 2018. Vol. 161. Pp. 330-339.
[83] Jian Qiu , Shaopeng Wu, A. A. A. Molenaar Investigating the Self Healing Capability of Bituminous Binders. *Road Materials and Pavement Design*. 2009. Vol. 10. Pp. 81-94.
[84] Norambuena-Contreras J., Liu Q., Zhang L., Wu S., Yalcin E., Garcia A. Influence of encapsulated sunflower. *Materials and Structures*. 2019. Vol. 52. Issue 4. 78.
[85] Xu S., Tabaković A., Liu X., Palin D., Schlangen E. Optimization of the calcium alginate capsules for self-healing asphalt. *Materials*. 2019. Vol. 12. Issue. 1. 168.
[86] Tabaković A., Schuyffel L., Karač A., Schlangen E. An evaluation of the efficiency of compartmented alginate fibres encapsulating a rejuvenator as an asphalt pavement healing system. *Applied Sciences*. 2017. Vol. 7. Issue 7. 647.
[87] Shu B., Bao S., Wu S., Dong L., Li C., Yang X., Norambuena-Contreras J., Liu Q., Wang Q. Synthesis and effect of encapsulating rejuvenator fiber on the performance of asphalt mixture. 2019. *Materials*. Vol. 12. Issue 8. 1266.

Acknowledgment(s)
This work was financially supported by the Russian Science Foundation (project # 19-79-00262).

IOP Conf. Series: Materials Science and Engineering **855** (2020) 012011 doi:10.1088/1757-899X/855/1/012011

A Model for FGM/SMA Bilayer Beams Accounting for Asymmetric SMA Behavior

W Zaki* and N V Viet

Khalifa University of Science and Technology, P.O. Box 127788, Abu Dhabi, UAE

wael.zaki@ku.ac.ae

Abstract. A new model is proposed for bilayer cantilever beams consisting of shape memory alloy (SMA) and functionally graded material (FGM) layers and subjected to loading at the tip. The model accounts for tensile-compressive SMA behavior through the use of an appropriate set of constitutive relations and considers the deformation of the beam within the frame of the Timoshenko beam theory. The derivation of the model proceeds by first determining the correct sequence in which different solid phase structures develop within the SMA layer as the load is applied, resulting in the formation of distinct solid phase regions. The boundaries separating these regions are then located and used in deriving appropriate moment and shear force equations throughout a complete loading-unloading cycle. The model is capable of tracking the deviation of the neutral surface with respect to the mid-plane and the distribution of martensite in the beam as the load varies. The nonlinear variation of stress and strain in a cross section of the beam is also considered. Results of beam deflection and neutral surface deviation are presented and validated against finite element simulations of a 3D model of a beam consisting of a Nitinol and CNT-reinforced layers.

1. Introduction

Since their discovery in the early 1980s [1], FGMs have been used in various fields such as aerospace, electronics, medical technology, energy conversion, etc. This rapid development was motivated by the distinctive performance of FGMs, which cannot usually be achieved using monolithic or conventionally associated materials. Examples of such properties include high fracture toughness, low in-plane transverse stresses, enhanced thermal barrier performance, wear resistance, etc. [2]. In contrast, SMAs are well known for their superelasticity and shape memory effect [3], which are made possible by the ability of these alloys to undergo the so-called martensitic transformation between phases of different crystallographic symmetries. Combining FGMs and SMAs has therefore allowed the creation of "smart" FGM composites, combining the advantages of both types of materials and possibly paving the way for unprecedented engineering applications.

FGM/SMA laminates have recently attracted the interest of several research groups. In this regard, Viet et al. [3] proposed an analytical model for FGM/SMA laminates subjected to tip load based on the ZM model for SMAs and Timoshenko beam theory. Liu et al. [4] presented an analytical model for functionally graded SMA composite beams subjected to bending based on the Euler-Bernoulli theory. Sepiani et al. [5] developed a theoretical model for a FGM/SMA laminate beam subjected to arbitrary thermal loading and boundary conditions, etc. [6,7]. A limitation of the existing literature is the common disregard of tensile-compressive asymmetry in the response of the SMA constituent and of the behavior of the beam during unloading. Both these limitations are addressed here. Moreover, the deformation of the beam is described within the frame of the Timoshenko beam theory and the influence of the neutral axis deviation, as the load is varied, on the resulting moments and shear forces acting in the beam is fully considered.

2. Analytical model of the FGM/SMA laminated composite cantilever beam

The model is derived using the ZM constitutive theory to describe the behavior of the SMA component [8]. Multiple version of the ZM model have been developed over the years, collectively accounting for most of the unusual SMA effects [8-17].

2.1. *Beam structure and material properties of FGM*

The beam structure with the SMA layer bonded onto the FGM is shown in Figure 1.

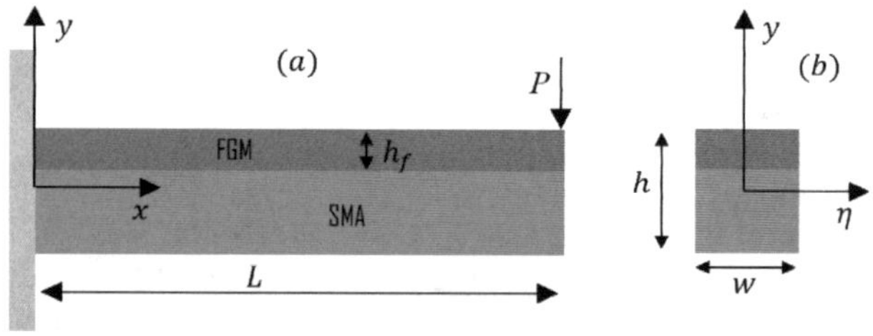

Figure 1. Geometric configuration and boundary conditions of the SMA/FGM beam. (a) Longitudinal view; (b) cross section view.

In the figure, w and h are the width and height of the composite, and h_f is the FGM thickness. h_s is the distance from the mid-plane to the interface between FGM and SMA layers computed as $h_s = h/2 - h_f$. Young's modulus, Poisson's ratio and density of FGM vary continuously along the thickness according to the volume fraction of its constituents [18]. The Young's modulus of the FGM, E_F, is given by

$$E_F = E_b + (E_t - E_b)\left(\frac{y + \Delta - h_s}{h/2 - h_s}\right)^n \tag{1}$$

where, E_t, E_b are Young's moduli of the inclusion and matrix, and y the distance of the neutral axis to the mid-plane. The FGM is assumed to consist of an epoxy matrix with Young's modulus $E_b = 3\ GPa$ with Carbon nanotube inclusions (CNTs) of Young's modulus $E_t = 1000\ GPa$.

2.2. *Analytical stress-strain relation for the SMA layer considering asymmetry*

Based on the loading functions for forward and reverse phase transformations accounting for the tensile-compressive stress asymmetry in the extended 3D ZM model for SMAs [19], the following reduced form of the loading functions is obtained:

$$F_{j\pm} = sign(\times)\left[\left(\frac{El_{MA} + P_{MA}}{2}\right)\sigma_{j\pm}^2 + \sigma_{j\pm}\varepsilon_{0\pm}\right] + \left\{a - b - sign(\times)\left(G + (\alpha - \beta)\left\{k_h\left(\left(\frac{3}{4}\right)^{3/2}\varepsilon_{0\pm}^3 - c_h\frac{\varepsilon_{0\pm}^3}{3.8}\right)\right\}^{2/3}\right)\right\}z$$

$$-sign(\times)\left(\frac{\beta}{2}\left\{k_h\left(\left(\frac{3}{4}\right)^{3/2}\varepsilon_{0\pm}^3 - c_h\frac{\varepsilon_{0\pm}^3}{3.8}\right)\right\}^{2/3} + C(T)\right) - a \tag{2}$$

where

$$sign(\times) = \begin{cases} 1 & \text{for loading} \\ -1 & \text{for unloading} \end{cases}$$

In Eq. (2), El_{MA} and P_{MA} are material parameters related to Young's moduli of austenite and martensite, ε_0 is the maximum orientation strain, G, a, b, α, β are material parameters that govern the size of the superelastic hysteresis loop and hardening during forward and reverse phase transformation, $C(T)$ is a temperature dependent energy term that governs the temperature dependence of the stress transformation thresholds of phase transformation, k_h and c_h are parameters that govern the extent of asymmetry between tension and compression.

The initial and final stresses for forward phase transformation, respectively denoted $\sigma_{\pm}^{Ms}$ and $\sigma_{\pm}^{Mf}$ can be found by substituting *z=0* and *z=1*, respectively, into Eq. (1) to obtain

$$\begin{cases} \sigma_{\pm}^{Ms} = \dfrac{-\varepsilon_{0\pm} + \sqrt{\varepsilon_{0\pm}^2 + 2(El_{MA} + P_{MA})(\beta\Gamma_{\pm}/2 + C(T) + a)}}{El_{MA} + P_{MA}} \\ \sigma_{\pm}^{Mf} = \dfrac{-\varepsilon_{0\pm} + \sqrt{\varepsilon_{0\pm}^2 - 2(El_{MA} + P_{MA})\{a - G - b - (\alpha - 3\beta/2)\Gamma_{\pm} + C(T) + a\}}}{El_{MA} + P_{MA}} \end{cases} \tag{3}$$

where $\Gamma_{\pm} = \left\{ k_h \left(\left(\dfrac{3}{4} \right)^{3/2} \varepsilon_{0\pm}^3 - c_h \dfrac{\varepsilon_{0\pm}^3}{3.8} \right) \right\}^{2/3}$.

Following the procedure in [20-25], the stress in the tensile and compressive regions during loading ($j = 1$) and unloading ($j = -1$) can be written in terms of strain as

$$\sigma_{j\pm} = \Pi_{j1\pm}^* + \Pi_{j2\pm}^* \varepsilon_{j\pm} + \Pi_{j3\pm}^* \varepsilon_{j\pm}^2 + \Pi_{j4\pm}^* \varepsilon_{j\pm}^3 + \Pi_{j5\pm}^* \varepsilon_{j\pm}^4 - \frac{\Pi_{j1\pm}}{3} \tag{4}$$

where the coefficients are constants obtained following the procedure described in [20] and given by the following expressions:

$$\Pi_{j1\pm} = \frac{2\varepsilon_{0\pm}\left(\dfrac{El_{MA} + P_{MA}}{2} + \dfrac{1}{E_M} - \dfrac{1}{E_A} \right)}{(El_{MA} + P_{MA})(1/E_M - 1/E_A)};$$

$$\Pi_{j2\pm} = 2\frac{\varepsilon_{0\pm}{}^2 - \dfrac{sign(\times)[a - b - (\alpha - \beta)\Gamma_{\pm}] - G}{E_A} - \left(\dfrac{\beta}{2}\Gamma_{\pm} + C(T) + sign(\times)a \right)\left(\dfrac{1}{E_M} - \dfrac{1}{E_A} \right)}{(El_{MA} + P_{MA})(1/E_M - 1/E_A)};$$

$$\Pi_{j3\pm} = 2\frac{sign(\times)(a - b) - G - (\alpha - \beta)\Gamma_{\pm}}{(El_{MA} + P_{MA})(1/E_M - 1/E_A)};$$

$$\Pi_{j4\pm} = -2\frac{\varepsilon_{0\pm}\left(\dfrac{\beta}{2}\Gamma_{\pm} + C(T) + sign(\times)a \right)}{(El_{MA} + P_{MA})(1/E_M - 1/E_A)}.$$

2.3. *Moment equation*

By assuming perfect bonding between the FGM and SMA layers, the strain is continuous across the FGM-SMA interface [26, 27]. The Timoshenko theory for the beam then gives

$$\begin{cases} \varepsilon = y d\theta / dx = yk \\ d\omega / dx = \gamma - \theta \end{cases} \tag{5}$$

where ω is the vertical displacement of the mid-plane, θ is the rotation of the normal to the mid-plane, $k = d\theta/dx$ is the curvature, γ is the transverse shear strain, and y is the vertical coordinate with respect to the neutral plane. During loading, the SMA beam can experience three stages including stages 1, 2 and 3 according to the number of material phases appearing in the beam. The details of these stages can be found in [20]. In this work, the notations $F_{xji\pm}$, $M_{ji\pm}$, $Q_{ji\pm}$, $\Delta_{ji\pm}$and $k_{ji\pm}$ as introduced to respectively indicate the force normal to the cross section, the bending moment, the shear force, the neutral axis deviation from the mid-plane in section i, and the curvature in section i. The width-normalized force, width-normalized moment and width-normalized shear force in section i are then defined as $\tilde{F}_{xji\pm} =$

IOP Conf. Series: Materials Science and Engineering **855** (2020) 012011 doi:10.1088/1757-899X/855/1/012011

$F_{xji\pm}/w$, $\widetilde{M}_{ji\pm} = M_{ji\pm}/w$, and $\tilde{Q}_{ji\pm} = Q_{ji\pm}/w$. A section, denoted s_i, is defined as a longitudinal portion of the beam containing cross sections with similarly ordered arrangements of solid phase regions (Figure 2).

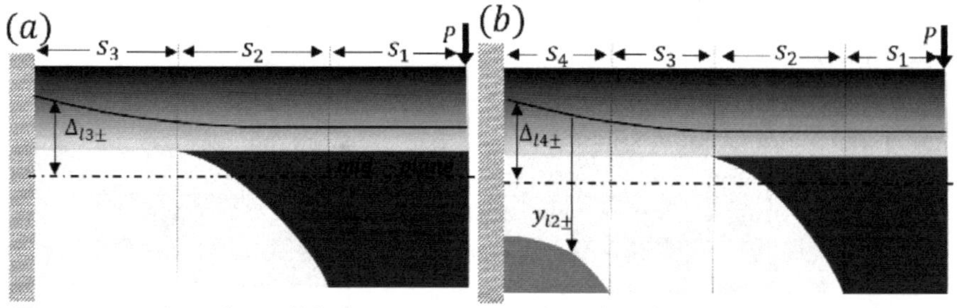

Figure 2. Two examples of possible beam structures in a longitudinal cross section view during loading. (a) Stage 2; (b) stage 3.

In Figure 2, austenite is in blue, martensite is in red, and regions of phase mixture are shown in yellow. The FGM material is in graded black.

Starting from an austenitic beam, which is not shown here, applying a monotonically increasing tip load results in increasing moment with highest magnitude at the lower left tip of the beam. Phase transformation therefore starts at that point and propagates to fill the yellow region in Figure 2(a). As the load increases further, austenite is fully transformed into martensite in the region highlighted in red in Figure 2(b).

The distance of the boundary between austenite and the mixed-phase region, and between the mixed-phase region and martensite to the neutral axis are given by

$$\begin{cases} y_{l1i\pm} = \dfrac{\sigma_{\pm}^{Ms}}{k_{li} E_A} = \dfrac{\rho_{1\pm}}{k_{li}} \\ y_{l2i\pm} = \dfrac{\sigma_{\pm}^{Mf} + \varepsilon_{0\pm} E_M}{k_{li} E_M} = \dfrac{\rho_{2\pm}}{k_{li}} \end{cases} \tag{6}$$

where $\rho_{1\pm} = \sigma_{\pm}^{Mf}/E_A$ and $\rho_{2\pm} = \{\sigma_{\pm}^{Mf} + \varepsilon_{0\pm} E_M\}/E_M$. The first equation in (6) are obtained by considering elastic behavior of austenite at the onset of forward martensite transformation, while the second equation is obtained considering the stress-strain relation in [20], specified to the case where the martensite volume fraction z is equal to 1, indicating complete forward phase transformation.

Due to the influence of neutral axis variation during forward phase transformation in the SMA layer, there is no unique beam phase structure. Thus, the geometric relation and axial force equilibrium are used to identify the correct beam phase structure by comparison among corresponding curvatures. For example, the curvature $\Delta_{2.3\pm}$ of a boundary cross section between s_2 and s_3 in Fig.2a based on axial force equilibrium and geometric relation during is obtained by solving the system

$$\begin{cases} \Delta_{2.3\pm} = h_s - \dfrac{\rho_{1\pm}}{k_{2.3}} \\ \sum \tilde{F}_{2.3x\pm} = \left\{ \left(\Pi_{l1\pm}^{*} - \Pi_{l1\pm}\right) y + \Pi_{l2\pm}^{*} \dfrac{k_{2.3\pm} y^2}{2} + \Pi_{l3\pm}^{*} \dfrac{k_{2.3\pm}^2 y^3}{3} + \Pi_{l4\pm}^{*} \dfrac{k_{2.3}^3 y^4}{4} + \Pi_{l5\pm}^{*} \dfrac{k_{2.3\pm}^4 y^5}{5} \right\} \Bigg|_{h/2-\Delta_{2.3\pm}}^{h_s-\Delta_{2.3\pm}} \\ + k_{2.3\pm} \left[E_b \dfrac{y^2}{2} + \dfrac{(E_t - E_b)(\psi_{1\pm} y + \psi_{2\pm})^{n+1} \{\psi_{1\pm}(n+1) y - \psi_{2\pm}\}}{\psi_{1\pm}^2 (n+1)(n+2)} \right] \Bigg|_{h_s-\Delta_{2.3\pm}}^{h/2-\Delta_{2.3\pm}} = 0 \end{cases} \tag{7}$$

where constants $\psi_{1\pm}$ and $\psi_{2\pm}$ are $\psi_{1\pm} = 1/(h/2 - h_s)$ and $\psi_{2\pm} = (\Delta_{ji\pm} - h_s)/(h/2 - h_s)$. The first equation in (7) expresses that the relation between the neutral axis deviation $\Delta_{2.3\pm}$ and the y coordinate of the interface between boundaries 2 and 3. The second equation expresses that the sum of resultant axial forces, obtained by integration of the normal stress over a cross sectional area, is equal to zero. Once the correct structure is determined, the analytical relation between moment, curvature and neutral axis deviation, and between shear force and shear strain can be obtained for each beam section. Without loss of generality, we demonstrate the derivation procedure of such relation for section s_4 of case 4. Indeed, the equilibrium of forces and moments for particular cross section of s_4 is written as follows:

$$\begin{cases} \tilde{F}_{xl4\pm} = \int_{-h/2-\Delta_{l4\pm}}^{y_{l24\pm}} E_M\left(k_{l4\pm}y - \varepsilon_{0\pm}\right)dy + \int_{y_{l24\pm}}^{h_s-\Delta_{l4\pm}} \sigma_{l\pm}dy + \int_{h_s-\Delta_{l4\pm}}^{h/2-\Delta_{l4\pm}} k_{l4\pm}E_F\, y dy = 0 \\ \tilde{M}_{l4\pm} = \int_{-h/2-\Delta_{l4\pm}}^{y_{l24\pm}} E_M\left(k_{l4\pm}y - \varepsilon_{0\pm}\right) y dy + \int_{y_{l24\pm}}^{h_s-\Delta_{l4\pm}} \sigma_{l\pm} y dy + \int_{h_s-\Delta_{l4\pm}}^{h/2-\Delta_{l4\pm}} k_{l4\pm}E_F\, y^2 dy = 0 \end{cases} \tag{8}$$

where k_{l4} is the curvature and Δ_{l4} the deviation of the neutral axis through section s_4. Substituting the expressions of the stresses from Eq. (4), and the values of boundary distances in Eq. (6) into Eq. (8), the relation for section s_4 of case 4 is written as

$$\begin{cases} \tilde{F}_{xl4\pm} = \int_{-h/2-\Delta_{l4\pm}}^{y_{l24\pm}} E_M\left(k_{l4\pm}y - \varepsilon_{0\pm}\right)dy + \int_{y_{l24\pm}}^{h_s-\Delta_{l4\pm}} \sigma_{l\pm}dy + \int_{h_s-\Delta_{l4\pm}}^{h/2-\Delta_{l4\pm}} k_{l4\pm}E_F\, y dy = 0 \\ \tilde{M}_{l4\pm} = \int_{-h/2-\Delta_{l4\pm}}^{y_{l24\pm}} E_M\left(k_{l4\pm}y - \varepsilon_{0\pm}\right) y dy + \int_{y_{l24\pm}}^{h_s-\Delta_{l4\pm}} \sigma_{l\pm} y dy + \int_{h_s-\Delta_{l4\pm}}^{h/2-\Delta_{l4\pm}} k_{l4\pm}E_F\, y^2 dy = 0 \end{cases} \tag{9}$$

The formulation of the volume fraction z in terms of the coordinate y measured with respect to the neutral axis within the transformed region is considered to be linear following [20]:

$$z_{l4\pm} = a_{l4\pm}y + b_{l4\pm} \tag{10}$$

where z is equal to 1 at the boundary between the mixed-phase and martensitic regions. The coefficients in Eq. (10) are $a_{l4\pm} = (1 - z_{4\mathrm{int}\pm})/(\rho_{2\pm}/k_{l4\pm} - h_s + \Delta_{l4\pm})$ and $b_{l4\pm} = 1 - a_{l4\pm}\rho_{2\pm}/k_{l4\pm}$.
Following [20], the shear force in section s_4 is obtained based on Eq. (10) as

$$\begin{aligned} \tilde{Q}_{l4\pm} = K_s G_M G_A \Bigg[& \frac{\ln\{G_M + b_{l4\pm}(G_A - G_M)y\}}{a_{l4\pm}(G_A - G_M)}\Bigg|_{\frac{\rho_{2\pm}}{k_{l4\pm}}}^{h_s-\Delta_{l4\pm}} \gamma_{l4\pm} - \gamma_{0\pm}\frac{a_{l4\pm}(G_A - G_M)y - G_M\ln\{G_M(a_{l4\pm}y + b_{l4\pm} - 1) - G_A(a_{l4\pm}y + b_{l4\pm})\}}{a_{l4\pm}(G_A - G_M)^2}\Bigg|_{\frac{\rho_{2\pm}}{k_{l4\pm}}}^{h_s-\Delta_{l4\pm}} \Bigg] \\ & + K_s G_M\left(\gamma_{l4\pm} - \gamma_{0\pm}\right)\Big|_{-h/2-\Delta_{l4\pm}}^{\frac{\rho_{2\pm}}{k_{l4\pm}}} + K_s\gamma_{l4\pm}\left[G_b\frac{y^2}{2} + \frac{(G_t - G_b)(\psi_{1\pm}y + \psi_{2\pm})^{n+1}\{\psi_{1\pm}(n+1)y - \psi_{2\pm}\}}{\psi_{1\pm}^2(n+1)(n+2)}\right]\Bigg|_{h_s-\Delta_{l4\pm}}^{\frac{h}{2}-\Delta_{l4\pm}} \end{aligned} \tag{11}$$

where the magnitude γ_0 of the transformation shear strain is $\gamma_{0+} = \sqrt{3}\varepsilon_{0+}/2$; K_s is the Timoshenko shear correction factor for rectangular sections $K_s = 5/6$; and G_F, G_b and G_t are the shear moduli of the FGM layer corresponding to Young's moduli E_F, E_b and E_t, respectively.

3. Finite Element Analysis and Validation of the Model

Finite element analysis of a 3D model of the beam is carried out using a (UMAT) that implements the ZM model in ABAQUS to simulate the behavior of the SMA layer. The dimensions and material properties are identical to those used in solving the analytical model. The bottom-up mesh tool in ABAQUS is used to mesh the FGM layer into 26 thin layers, each assigned a different set of material properties based on Eq. (1). The computation of the deflection of the beam using the analytical model is carried out in two steps: a curvature function is determined from fitting discrete curvature data along the beam length using a trust region algorithm in MATLAB [32], the deflection is then found by double integration [20].

4. Results and Discussions

The input parameters of the beam used in the analytical method solution and 3D FEA are provided in Table 1.

Table 1. Material properties and dimensions of the laminated beam.

Material properties				Dimensions	
E_A	61500 MPa	a	6.8920 MPa	c_h	1.299
E_M	24000 MPa	b	6.9091 MPa	w	0.5 mm
v	0.3	k_h	3.29	L	50 mm
	2750 MPa	G	4.6556 MPa	h	3 mm
A_f^0	40 ^{0}C		5500 MPa	h_f	1 mm
	0.4381MPa/^{0}C	k_T	2.492 MPa		
ε_{0+}	4 %	ε_{0-}	-2.5 %		

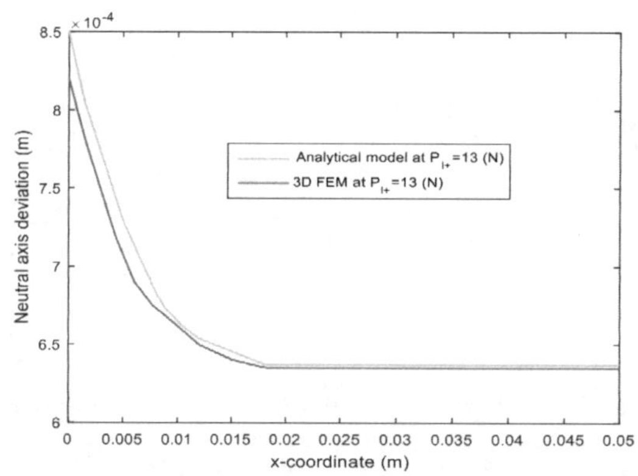

Figure 3. Position of the neutral axis along the length of beam for an upward load $P_{l+} = 13\ N$, for T=45 C and n=4.

The position of the neutral axis along the length of beam for an upward load $P_{l+} = 13\ N$, for T=45 C and n=4 calculated using FEA and the analytical model is shown in Figure 3.

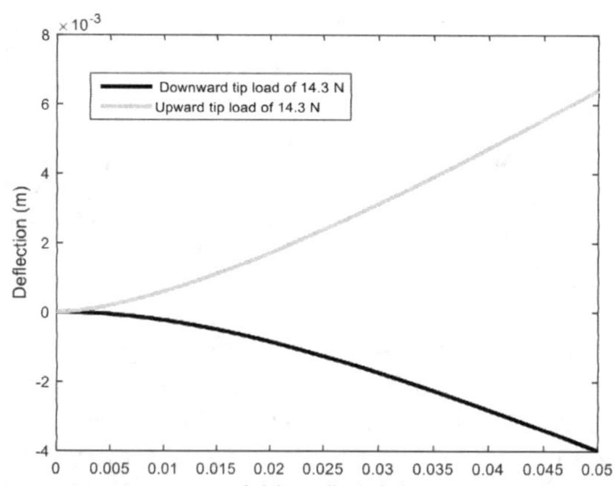

Figure 4. Beam deflection under upward and downward loads $|P_{l\pm}| = 11\ N$ for $n = 4$ and T=45 ^{0}C.

A good agreement is obtained with a maximum percent difference between the two methods of approximately 4% in terms of axial deviation from the mid-plane at the clamped face.

Figure 4 demonstrates the shape of the beam subjected to upward and downward loads of magnitude $|P_{l\pm}| = 11\ N$ for $n = 4$ and $T = 45\ ^0$C based on analytical model.

The difference in the deflected shape is due to different distributions of solid phase regions in each case.

5. Conclusions

In this work, an analytical model was derived for a FGM/SMA bilayer cantilever beam subjected to tip load. The model considers asymmetric behavior of the SMA in tension and compression and accounts for the variation of the position of the neutral axis of the beam with the applied load through a complete loading cycle. The results in terms of beam deflection profile and neutral axis deviation from the mid-plane based on the analytical model are validated against 3D FEA. The results indicate that under the same tip load magnitude, the overall stiffness of the FGM/SMA beam for a given load magnitude depends on the direction of the applied load because of asymmetric geometry and SMA material response.

6. References

[1] M. Naebe, and K. Shirvanimoghaddam. "Functionally graded materials: A review of fabrication and properties," App. Mater. Today. 5, pp. 223–245, 2016.

[2] Functionally Graded Materials (FGM) and Their Production Method, Azom.com. 2002-08-22.

[3] N.V. Viet, W. Zaki, and R. Umer. "Analytical model of functionally graded material/shape memory alloy composite cantilever beam under bending," Compos. Struct. 203, pp. 764-776, 2018

[4] L. Bingfei, G. Dui, and S. Yang. "On the transformation behavior of functionally graded SMA composites subjected to thermal loading," Eur. J. Mech. A/Solids. 40: 139e147, 2013.

[5] H. Sepiani, F. Ebrahimi, and H. Karimipour H. "A mathematical model for smart functionally graded beam integrated with shape memory alloy actuators," J. Mech. Sci. Tech. 23, pp.3179-3190, 2009.

[6] L. Xue, G. Dui, B. Liu, and J. Zhang. "Theoretical Analysis of a Functionally Graded Shape Memory Alloy Plate Under Graded Temperature Loading," Mech. Adv. Mater. Struct. 2016.

[7] H. Asadi, A.H. Akbarzadeh, and Q. Wang. "Nonlinear thermo-inertial instability of functionally graded shape memory alloy sandwich plates," Compos. Struct. 120, pp. 496–508, 2015.

[8] W. Zaki, and Z. Moumni. ''A three-dimensional model of the thermomechanical behavior of shape memory alloys,'' J. Mech. Phys. Solids. 55, pp. 2455–2490, 2007.

[9] C. Morin, Z. Moumni, and W. Zaki. "Direct numerical determination of the asymptotic cyclic behavior of pseudoelastic shape memory structures," J. Eng. Mech. 137(7), pp. 497-503, 2011.

[10] C. Cissé, W. Zaki, X. Gu, and T.B. Zineb. ''A nonlinear 3D model for iron-based shape memory alloys considering different thermomechanical properties for austenite and martensite and coupling between transformation and plasticity,'' Mech. Mater. 107, pp.1-21, 2017.

[11] Z. Moumni, W. Zaki, Q.S. Nguyen, and W. Zhang. ''Modeling of Materials Capable of Solid-Solid Phase Transformation. Application to the Analytical Solution of the Semi-Infinite Mode III Crack Problem in a Phase-Changing Solid,'' Int. J. Non-Linear Mech. 69, pp. 146-156, 2015.

[12] W. Zaki. "An approach to modeling tensile-compressive asymmetry for martensitic shape memory alloys," Smart. Mater. Struct. 19(2), 2010.

[13] W. Zaki, Z. Moumni, and C. Morin. ''Modeling Tensile-Compressive Asymmetry for Superelastic Shape Memory Alloys,'' Mech. Adv. Mater. Struct. 18 (7), pp. 559-564, 2011.

[14] J. Wang, Z. Moumni, W. Zhang, Y. Xu, and W. Zaki. ''A 3D Finite-Strain-Based Constitutive Model for Shape Memory Alloys Accounting for Thermomechanical Coupling and Martensite Reorientation''. Smart. Mater. Struct. 26 (6), 65006, 2017.

[15] J. Wang, Z. Moumni, W. Zhang, and W. Zaki. "A thermomechanically coupled finite deformation constitutive model for shape memory alloys based on Hencky strain," Int. J. Eng. Sci. 117, pp. 51-77, 2017.

[16] X. Gu, W. Zaki, C. Morin, Z. Moumni, and W. Zhang.''Time Integration and Assessment of a Model for Shape Memory Alloys Considering Multiaxial Nonproportional Loading Cases,'' Int. J. Solids. Struct. 54, pp. 82–99, 2015.
[17] W. Zaki. "Time integration of a model for martensite detwinning and reorientation under nonproportional loading using Lagrange multipliers," Int. J. Solids. Struct. 49(21), pp. 2951–2961, 2012.
[18] G. Bao, and L. Wang. "Multiple cracking in functionally graded ceramic/metal coatings," Int. J. Solids. Struct. 32, pp. 2853–87, 1995.
[19] W. Zaki, Z. Moumni, and C. Morin. "Modeling Tensile-Compressive Asymmetry for Superelastic Shape Memory Alloys," Mech. Adv. Mater. Struct. 18 (7), 2011.
[20] N.V. Viet, W. Zaki, and R. Umer. "Bending models for superelastic shape memory alloy laminated composite cantilever beams with elastic core layer," Compos. Part B, 2018.
[21] N.V. Viet, W. Zaki, R. Umer, and C. Cisse. "Analytical model for the torsional response of superelastic shape memory alloy circular sections subjected to a loading-unloading cycle," Inter. J. Solid. Struct. 156, pp. 49-60, 2019.
[22] N.V. Viet, W. Zaki, and R. Umer. "Analytical Model for a Superelastic Timoshenko Shape Memory Alloy Beam Subjected to a Loading-Unloading Cycle," J. Intell. Mater. Syst. Struct. 2018.
[23] N.V. Viet, and W. Zaki. "Analytical investigation of the behavior of concrete beams reinforced with multiple circular superelastic shape memory alloy bars," Compos. Struct. 210, pp. 958-970, 2019.
[24] N.V. Viet, W. Zaki, and R. Umer. "Bending theory for laminated composite cantilever beams with multiple embedded shape memory alloy layers," J. Intell. Mater. Syst. Struct. In press.
[25] N.V. Viet, W. Zaki, and R. Umer. "Analytical Model for a Superelastic SMA Beam," ASME 2017 SMASIS, 2017.
[26] N.V. Viet, Q. Wang, and W.S. Kuo. "A studying on load transfer in carbon nanotube/epoxy composites under tension," J. Model. Mech. Mater. 2017.
[27] N.V. Viet, W. Zaki, and R. Umer. "Interlaminar Shear Stress Function for Adhesively Bonded Multi-Layer Metal Laminates,'' *Int. J. Adhes. Adhes. 82, 14-20, 2018.*

Acknowledgement
Dr. Wael Zaki would like to acknowledge the financial support of Khalifa University through CIRA grant no. CIRA2019-024.

IOP Conf. Series: Materials Science and Engineering **855** (2020) 012012 doi:10.1088/1757-899X/855/1/012012

Smart Material Serves Other Smart Materials: Comparison of Measurement of CO_2 Concentration by Optical and Optoacoustic Methods

Dostál Michal[1], Suchánek Jan[1], Zelinger Zdeněk[1], Wild Jan[2], Bitala Petr[2], Dudáček Aleš[2], Válek Václav[2], Klečka Vít[2], Kučera Petr[2], Nevrlý Václav[3]

[1]J. Heyrovský Institute of Physical Chemistry, Czech Academy of Sciences of Czech Republic, Prague, Czech Republic
[2]Faculty of Safety Engineering, VŠB – Technical University of Ostrava, Ostrava, Czech Republic
[3]Gaudi (Democratic Free School), Příbor, Czech Republic

e-mail: zdenek.zelinger@jh-inst.cas.cz

Abstract. In this study, we compare the sensitivity of the Wavelength Modulation Spectroscopy (WMS) to the Quartz enhanced photoacoustic spectroscopy (QEPAS) method on a sample of CO_2 gas at various concentrations. We show the advantages of the QEPAS method, the essential part of which is a crystalline segment made of smart material.

1. Introduction

One of the urgent problems of present civilization is content of carbon dioxide in the air. Within this context, the importance of smarts materials, which can serve as the basis for various devices to reduce emissions of this gas, such as scrubbers or catalysts, is increasingly important [1–5].

No less important role is played by various types of detection methods of carbon dioxide. Besides classical optical methods [6–8], photoacoustic spectroscopy, has some advantages over the optical methods.

In this study, we focus on comparing the optical method WMS [9] with QEPAS method where the function of sensor is performed by acoustic detector – Quartz Tuning Fork (QTF). Modified commercial device (KB Spektrolap TM) was used as optoacoustic spectrometer, which key part (fork of Mrs. Vice-dean) is also made from smart material and is used thanks to its specific material properties.

Both employed spectroscopy methods are based on wave modulation technique. Main advantages of the WMS are high sensitivity, fast reaction, simple construction and maintenance, noninvasive measurement and low operating cost [10]. It is possible to measure trace amounts of gases up to ppm in Near Infrared Range (NIR) with weak "overtones" and combination spectral bands. Often several different substances can be selectively measured using a single laser diode. Laser diodes typically have narrow emission lines allowing the measurement of narrow separated rotational-vibrational bands of absorbing substances.

The Quartz Enhanced Photoacoustic Absorption Spectroscopy (QEPAS) [11] method is based on the sensing of acoustic waves in a confined space using WMS principles. Acoustic waves are generated by

rotational-vibrational relaxation of molecules due to modulated infrared laser radiation. The acoustic signal in a cell can be easily measured using the Quartz Tuning Fork (QTF) with a high quality factor Q [11–13]. These QTFs are commonly used in electrical engineering as crystal oscillators and are made of silicon crystal cut, making them smart. The signal generated by QTF is amplified using a high input resistance transimpedance preamplifier. This method is described in Kosterev [14, 15] or Zheng [16], where they use parallel crystal oscillations.
The laser is focused using optical elements between the QTF tips and modulated at $f/2$ where f is the QTF resonant frequency. This arrangement creates acoustic waves with a corresponding frequency in the cell, which influence the oscillation of the crystal.

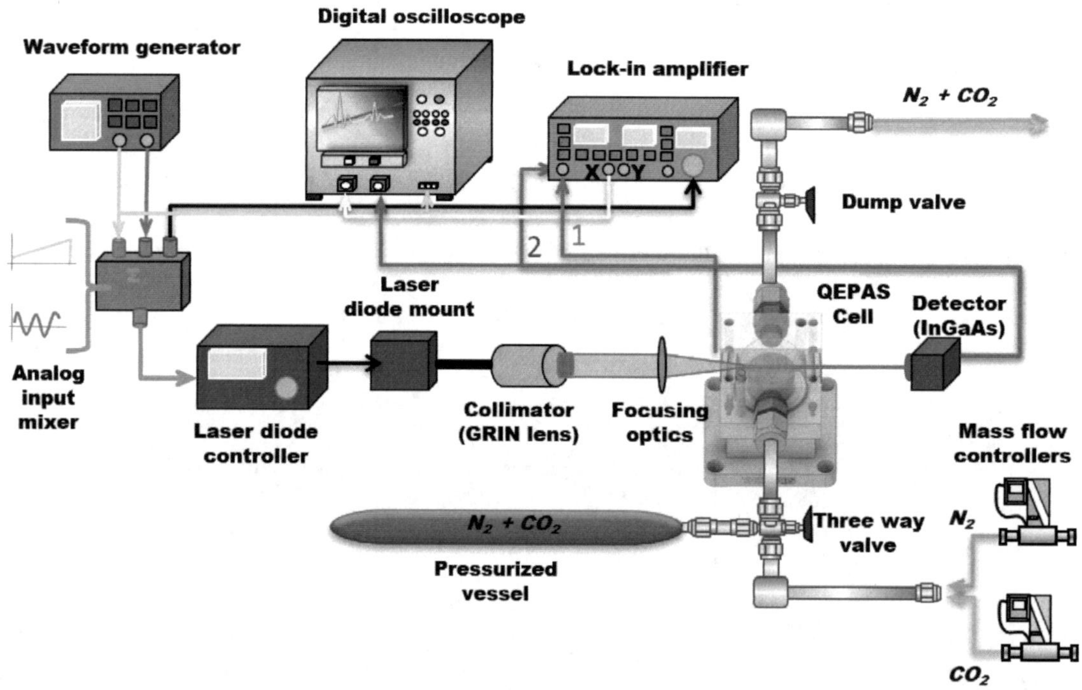

Figure 1. Experimental Setup, case 1 is the wiring for the QEPAS method, case 2 is the wiring for the optical WMS method.

2. Experimental

2.1. QEPAS method

The experimental set-up enabled measurement by both QEPAS and WMS methods on the same sample (**Figure 1**). A QTF with a resonant frequency f = 12454.5 Hz and a quality factor Q>12000 was used as an acoustic signal detection element. A diode laser (Eblana - EP 1573-0-DM-B01-FA) with a wavelength of about 1573 nm (corresponding to the wavenumber range 6358-6362 cm^{-1}) was used as the infrared radiation source. CO_2 absorption lines occur in this region. The wavelength of the infrared radiation emitted by the laser diode is dependent on the electrical current flowing through the laser diode and its temperature (kept at 28 °C). By changing both of these parameters, the wavelength of the emitted radiation can be varied continuously over a narrow wavelength range. The current and temperature stabilization of the diode was ensured by the control unit (Throlabs ITC 4001). The laser wavelength was externally modulated using a two-channel generator (LeCroy Wavestation 2012) with a slow asymmetric triangular signal at 20 mHz and a fast sinusoidal signal (6227.25 Hz) corresponding to $f/2$ with an amplitude of 100 mV. The laser beam was directed through a fiber optic to a QEPAS cell (Thorlabs ADM01). An InGaAs photodetector (Thorlabs PDA10T-EC) was placed behind the QEPAS cell to measure the intensity of radiation passing through the system. The

photodetector was connected to an oscilloscope. The signal measured by the QEPAS detector was then demodulated using a phase-sensitive lock-in amplifier (Standford Research AS830DSP). The lock-in amplifier was set to demodulate the second harmonic component of the signal corresponding to a frequency f (sensitivity 200 mV, time constant 300 ms). The amplified sine signal from the function generators, as described above, was used as a reference signal for demodulation. The output of the Lock-in amplifier is the X and Y components of the demodulated signal R.
In our case, the phase shift of the Lock-in amplifier has been set so that all output is concentrated on the X component of the signal. This output was connected to the first oscilloscope channel (LeCroy wavesurfer 42XS) through which the data was recorded. The output from the optical detector was connected to the second channel.

2.2. *WMS method*
For the optical WMS measurement, a photodetector located behind the photoacoustic cell was used to measure the change in intensity of the laser beam after passing through the cell. The output of the detector was connected to the Lock-in input of the amplifier, which was set with the same parameters as the QEPAS method. Also, the laser modulation frequency and modulation depth were set with the same parameters as the previous method. The output of the lock-in amplifier was connected to the first channel of the oscilloscope and the output from the photodetector was connected to the second channel.

2.3. *Gas sample preparation*
The defined concentration of CO_2 in N_2 was prepared using a flowmeter system (Bronkhorst F201-CV series). The selected concentration was first mixed into an expansion vessel and then filled into a cell. After filling with gas, the cell was closed.

3. Results and discussion
Figure 2 shows measured CO_2 spectra by WMS and QEPAS at two defined concentrations. From the measured data we can conclude that the QEPAS method achieves much higher sensitivity under the same conditions (absorption path).

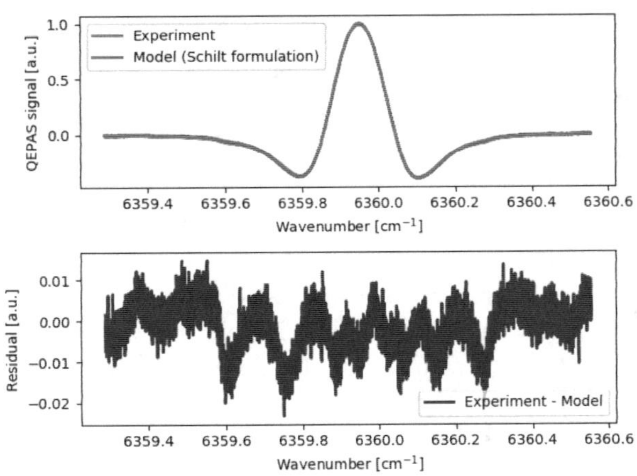

Figure 2. Measured CO_2 absorption spectra by QEPAS compared to WMS at two different concentrations. Case a) 100% CO_2, case b) 50% CO_2.

The most significant difference between the two methods is seen in case b) at a CO_2 concentration of 50%. While QEPAS still reliably detects this amount, the WMS optical method is at its threshold. Note that QEPAS can also detect units of percent of CO_2 concentration. Case a) clearly shows how optical components bring undesirable interference into spectroscopic information. In contrast, QEPAS does not show artifacts of this kind.

The measured data were analyzed using relevant mathematical models of spectral lines. The middle absorption line at 6369.5cm^{-1} was chosen for the modeling. The analysis shows that under given experimental conditions the measured data differ by ± 2% from the model case by the QEPAS method (**Figure 3**). From this we can conclude that the QEPAS method has a very good signal to noise ratio. Note that the measured data were processed using the algorithm described in the articles [17, 18].

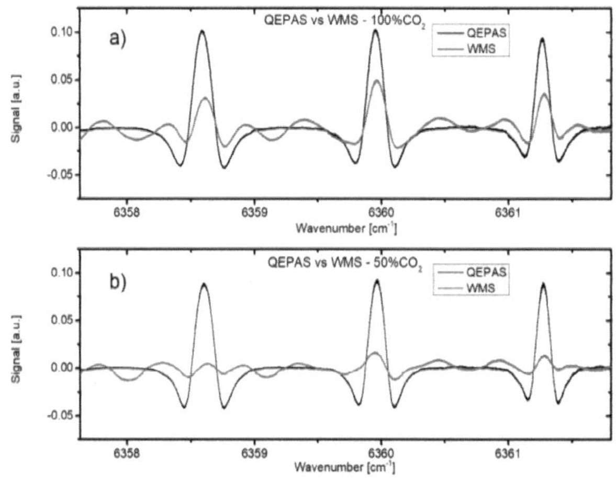

Figure 3. Spectral profile of R (16) absorption line of carbon dioxide and results of line-shape.

4. Conclusion

The experiment was focused on testing the possibility of using a crystalline QTF oscillator for gas detection using the QEPAS method compared to the optical WMS method. Under comparable conditions, e.g. the same absorption pathway, QEPAS has proven to be a much more sensitive and flexible CO_2 detection tool.

5. References

[1] Wurzbacher JA, Gebald C, Steinfeld A (2011) Separation of CO2 from air by temperature-vacuum swing adsorption using diamine-functionalized silica gel. Energy Environ Sci 4:3584. https://doi.org/10.1039/c1ee01681d

[2] Shi B, Xu W, Wu E, et al (2018) Novel design of integrated gasification combined cycle (IGCC) power plants with CO2 capture. J Clean Prod 195:176–186. https://doi.org/10.1016/J.JCLEPRO.2018.05.152

[3] Wang Y, Yang RT (2019) Chemical Liquid Deposition Modified 4A Zeolite as a Size-Selective Adsorbent for Methane Upgrading, CO $_2$ Capture and Air Separation. ACS Sustain Chem Eng 7:3301–3308. https://doi.org/10.1021/acssuschemeng.8b05339

[4] Keith DW, Ha-Duong M, Stolaroff JK (2006) Climate Strategy with Co2 Capture from the Air. Clim Change 74:17–45. https://doi.org/10.1007/s10584-005-9026-x

[5] Jansen D, Oudhuis ABJ, van Veen HM (1992) CO2 reduction potential of future coal gasification based power generation technologies. Energy Convers Manag 33:365–372. https://doi.org/10.1016/0196-8904(92)90032-R

[6] Schiff HI, I. H (1986) Tunable-diode laser absorption spectrometry. NASA Langley Res Cent Futur Dir H sub x O sub y Detect p 58-59 (SEE N87-15528 07-46)

[7] Bowling DR, Sargent SD, Tanner BD, Ehleringer JR (2003) Tunable diode laser absorption spectroscopy for stable isotope studies of ecosystem–atmosphere CO2 exchange. Agric For Meteorol 118:1–19. https://doi.org/10.1016/S0168-1923(03)00074-1

[8] Pogány A, Werhahn O, Wagner S, Ebert V (2015) Development and Metrological Characterization of a Tunable Diode Laser Absorption Spectroscopy (TDLAS) Spectrometer for Simultaneous Absolute Measurement of Carbon Dioxide and Water Vapor. Appl Spectrosc Vol 69, Issue 2, pp 257-268 69:257–268

[9] Moses EI, Tang CL (1977) High-sensitivity laser wavelength-modulation spectroscopy. Opt Lett 1:115. https://doi.org/10.1364/OL.1.000115

[10] Hodgkinson J, Tatam RP (2013) Optical gas sensing: a review. Meas Sci Technol 24:012004. https://doi.org/10.1088/0957-0233/24/1/012004

[11] Patimisco P, Scamarcio G, Tittel FK, Spagnolo V (2014) Quartz-enhanced photoacoustic spectroscopy: a review. Sensors (Basel) 14:6165–6206. https://doi.org/10.3390/s140406165

[12] Patimisco P, Sampaolo A, Dong L, et al (2016) Analysis of the electro-elastic properties of custom quartz tuning forks for optoacoustic gas sensing. Sensors Actuators B Chem 227:539–546. https://doi.org/10.1016/J.SNB.2015.12.096

[13] Spagnolo V, Patimisco P, Borri S, et al (2014) Terahertz quartz enhanced photo-acoustic gas sensor employing a quantum cascade laser source

[14] Kosterev AA, Bakhirkin YA, Curl RF, Tittel FK (2002) Quartz-enhanced photoacoustic spectroscopy. Opt Lett 27:1902. https://doi.org/10.1364/OL.27.001902

[15] Petra N, Zweck J, Kosterev AA, et al (2009) Theoretical analysis of a quartz-enhanced photoacoustic spectroscopy sensor. Appl Phys B 94:673–680. https://doi.org/10.1007/s00340-009-3379-1

[16] Zheng H, Dong L, Yin X, et al (2015) Ppb-level QEPAS NO2 sensor by use of electrical modulation cancellation method with a high power blue LED. Sensors Actuators B Chem 208:173–179. https://doi.org/10.1016/j.snb.2014.11.015

[17] Nevrlý V, Klečka V, Vašinek M, et al (2018) A new line-shape asymmetry model for wavelength modulation spectroscopy in gaseous flows. Meas Sci Rev 18:158–161. https://doi.org/10.1515/msr-2018-0022

[18] Nevrlý V, Bitala P, Klečka V, et al (2019) Analysis of Wavelength Modulation Spectra for Determination of OH Radical Concentration in an Atmospheric Pressure Laminar Premixed Flames. J Phys Conf Ser 1289:012046. https://doi.org/10.1088/1742-6596/1289/1/012046

Acknowledgment

The authors are grateful for the financial support from the Czech Science Foundation, Project No. 17-05167S, from the project No. VI20192022127 funded by the Ministry of the Interior of the Czech Republic and this work is a part of a research series funded by ERDF/ ESF “Centre of Advanced Applied Sciences” (No. CZ.02.1.01/0.0/0.0/16019/ 0000778). VSB-Technical University of Ostrava (Faculty of Safety Engineering) is also acknowledged for support via project No. SP2018/179 and project No. SP2019/158

AUTHOR INDEX